MEDICAL
IMAGE PROCESSING

MEDICAL IMAGE PROCESSING

Advanced Fuzzy Set Theoretic Techniques

TAMALIKA CHAIRA

CRC Press
Taylor & Francis Group
Boca Raton London New York

CRC Press is an imprint of the
Taylor & Francis Group, an **informa** business

CRC Press
Taylor & Francis Group
6000 Broken Sound Parkway NW, Suite 300
Boca Raton, FL 33487-2742

First issued in paperback 2020

© 2015 by Taylor & Francis Group, LLC
CRC Press is an imprint of Taylor & Francis Group, an Informa business

No claim to original U.S. Government works

ISBN 13: 978-0-367-57596-0 (pbk)
ISBN 13: 978-1-4987-0045-0 (hbk)

Library of Congress Cataloging-in-Publication Data

Chaira, Tamalika.
    Medical image processing : advanced fuzzy set theoretic techniques / Tamalika Chaira.
        pages cm
    Includes bibliographical references and index.
    ISBN 978-1-4987-0045-0 (hardback)
    1. Diagnostic imaging--Mathematics. 2. Fuzzy systems in medicine. 3. Cluster set theory. I. Title.

RC78.7.D53C456 2015
616.07'543--dc23   2014040185

Visit the Taylor & Francis Web site at
http://www.taylorandfrancis.com

and the CRC Press Web site at
http://www.crcpress.com

To my parents

Barid Baran Chaira and Puspa Chaira

and

daughter Shruti De

Contents

Preface...xiii
Organization of the Book...xv
Author...xvii

1. Intuitionistic Fuzzy Set and Type II Fuzzy Set.......................................1
  1.1   Introduction...1
  1.2   Intuitionistic Fuzzy Set..2
  1.3   Some Operations on Intuitionistic Fuzzy Sets 3
  1.4   Fuzzy Complement and Intuitionistic Fuzzy Generator 5
  1.5   Intuitionistic Fuzzy Generator..8
  1.6   Intuitionistic Fuzzy Relations 10
  1.7   Composition of Intuitionistic Fuzzy Relation
        (Supremum–Infimum)...11
  1.8   Composition of Intuitionistic Fuzzy Relation Using Fuzzy
        t-Norm and t-Conorm 12
        1.8.1 t-Norm 12
        1.8.2 t-Conorm 12
        1.8.3   Reflexive Property 14
        1.8.4   Symmetric Property 15
        1.8.5   Transitive Property...15
  1.9   Interval-Valued Intuitionistic Fuzzy Set....................................16
  1.10 Type II Fuzzy Set..18
  1.11 Summary..20
  References 20

2. Medical Image Processing...23
  2.1   Introduction...23
        2.1.1   Image Contrast Enhancement..23
        2.1.2   Image Segmentation...24
        2.1.3   Boundary Detection 24
        2.1.4   Morphology 24
  2.2   Image Registration..24
  2.3   Image Fusion..28
  2.4   Image Retrieval 28
  2.5   Fuzzy Processing of Medical Images...29
  2.6   Advanced Fuzzy Processing of Medical Images.........................31
        2.6.1   Intuitionistic Fuzzy Set..31
        2.6.2   Type II Fuzzy Set..32

2.7   Some Applications of Advanced Fuzzy Set in Medical
        Image Processing 33
2.8   Summary..36
References 37

**3. Fuzzy and Intuitionistic Fuzzy Operators with Application
   in Decision-Making**..39
3.1   Introduction..39
3.2   Fuzzy Operators...39
3.3   Fuzzy Operators Induced by Fuzzy *t*-Norm and *t*-Conorm........40
        3.3.1   *t*-Norm 41
        3.3.2   *t*-Conorm 42
        3.3.3   Negation..47
3.4   Fuzzy Aggregating Operators 47
        3.4.1   Weighted Averaging Operator................................48
        3.4.2   Ordered Weighted Averaging Operator...............48
3.5   Intuitionistic Fuzzy Weighted Averaging Operator 48
        3.5.1   Generalized Intuitionistic Fuzzy Weighted
                Averaging Operator...49
        3.5.2   Generalized Intuitionistic Fuzzy Ordered Weighted
                Averaging Operator..52
        3.5.3   Generalized Intuitionistic Fuzzy Hybrid Averaging
                Operator 54
3.6   Application of Intuitionistic Fuzzy Operators
        to Multi-Attribute Decision-Making 57
3.7   Intuitionistic Fuzzy Triangular Norms and Triangular
        Conorms 58
3.8   Summary..59
References 60

4. Similarity, Distance Measures and Entropy............................63
4.1   Introduction..63
4.2   Similarity Measure 64
        4.2.1   Similarity/Distance Measure................................64
        4.2.2   Distance Measures..65
4.3   Different Types of Distance and Similarity Measures 66
4.4   Intuitionistic Fuzzy Measure 70
4.5   Intuitionistic Fuzzy Information Measure......................73
4.6   Intuitionistic Fuzzy Entropy 74
        4.6.1   Different Types of Entropies 75
4.7   Entropy of Interval-Valued Intuitionistic Fuzzy Set......78
4.8   Similarity Measure and Distance Measures of *IVIFS* 79
4.9   Summary..80
References 81

5. Image Enhancement..83
  5.1   Introduction 83
  5.2   Fuzzy Image Contrast Enhancement 84
  5.3   Fuzzy Methods in Contrast Enhancement............................85
        5.3.1   Contrast Enhancement Using the Intensification
                    Operator 86
        5.3.2   Contrast Improvement Using Fuzzy Histogram
                    Hyperbolization 86
        5.3.3   Contrast Enhancement Using IF-THEN Rules 86
        5.3.4   Contrast Improvement Using the Fuzzy Expected Value....87
  5.4   Intuitionistic Fuzzy Enhancement Methods.........................88
        5.4.1   Entropy-Based Enhancement Methods 88
        5.4.2   Two-Dimensional Entropy–Based Intuitionistic
                    Fuzzy Enhancement (Method II)........................... 92
        5.4.3   Entropy-Based Enhancement Method by Chaira
                    (Method III)... 93
        5.4.4   Contrast Enhancement by Chaira (Method IV) 94
        5.4.5   Hesitancy Histogram Equalization.................................... 96
  5.5   Image Enhancement Using Type II Fuzzy Set..................... 99
        5.5.1   Type II Fuzzy Enhancement (Method I)........................... 99
        5.5.2   Enhancement Using Hamacher *t*-Conorm...................... 100
        5.5.3   Enhancement Using Type II Fuzzy Set............................ 100
  5.6   Introduction to MATLAB®.. 102
        5.6.1   Examples Using MATLAB.. 105
  5.7   Summary 107
  References 107

6. Thresholding of Medical Images 109
  6.1   Introduction 109
  6.2   Threshold Detection Methods 110
        6.2.1   Global Thresholding.. 110
        6.2.2   Iterative Thresholding... 111
        6.2.3   Optimal Thresholding 111
        6.2.4   Locally Adaptive Thresholding....................................... 111
        6.2.5   Locally Adaptive and Optimal Thresholding 113
                6.2.5.1   Chow and Kaneko Method.............................. 113
                6.2.5.2   Multispectral Thresholding............................. 114
  6.3   Fuzzy Methods.. 114
        6.3.1   Fuzzy Divergence Method 114
        6.3.2   Fuzzy Geometry Method 115
        6.3.3   Fuzzy Clustering Method... 116
  6.4   Intuitionistic Fuzzy Threshold Detection Methods.................... 117
        6.4.1   Intuitionistic Fuzzy Entropy–Based Method................... 118
        6.4.2   Intuitionistic Fuzzy Divergence–Based Method............. 121
                6.4.2.1   Intuitionistic Fuzzy Divergence Measure 123

6.5   Window-Based Thresholding 125
      6.5.1   Calculation of Membership Function 126
6.6   Thresholding Using Type II Fuzzy Set Theory 129
6.7   Segmentation Using Type II Fuzzy Set Theory 131
6.8   Segmenting Leucocyte Images in Blood Cells......................... 132
      6.8.1   Cauchy Distribution 132
6.9   Examples Using MATLAB® 136
      6.9.1   Intuitionistic Windowed Thresholding Method 136
6.10 Summary... 139
References 139

7. Clustering of Medical Images.. 141
7.1   Introduction 141
7.2   Fuzzy *c* Means Clustering 142
7.3   Hierarchical Clustering.. 144
7.4   Kernel Clustering 145
7.5   Kernel Clustering Methods 147
7.6   Intuitionistic Fuzzy *c* Means Clustering 153
7.7   Kernel-Based Intuitionistic Fuzzy Clustering 157
7.8   Colour Clustering... 161
      7.8.1   Colour Model... 162
7.9   Type II Fuzzy Clustering 165
7.10 Summary... 166
References 167

8. Edge Detection 169
8.1   Introduction 169
      8.1.1   Thresholding Method 170
      8.1.2   Hough Transform Method 171
      8.1.3   Boundary-Based Method.. 171
8.2   Fuzzy Methods.. 172
8.3   Intuitionistic Fuzzy Edge Detection Method............................ 173
      8.3.1   Template-Based Edge Detection 174
      8.3.2   Edge Detection Using the Median Filter 176
8.4   Fuzzy Edge Image Using Interval-Valued Fuzzy Relation 179
8.5   Construction of Enhanced Fuzzy Edge Using Type II
      Fuzzy Set 181
8.6   Accurate Edge Detection Technique 182
8.7   Implementation Using MATLAB® 186
      8.7.1   An Example to Find the Edge Image 186
      8.7.2   An Example to Find the Fuzzy Edge Image 188
8.8   Summary... 189
References 189

9. Fuzzy Mathematical Morphology 191
   9.1   Introduction 191
   9.2   Preliminaries on Morphology... 191
          9.2.1   Greyscale Mathematical Morphology.............................. 192
   9.3   Fuzzy Mathematical Morphology 193
          9.3.1   Different Definitions of Fuzzy Morphology 195
          9.3.2   Fuzzy Morphology Using Lukasiewicz Operator........... 196
          9.3.3   Fuzzy Morphology Using t-Norms and t-Conorms
                  by De Baets and Kerre and Bloch and Maitre 198
   9.4   Opening and Closing Operations..................................... 200
   9.5   Fuzzy Morphology in Image Processing......................... 205
          9.5.1   Edge Detection 205
          9.5.2   Intuitionistic Fuzzy Approach... 206
   9.6   Implementation in MATLAB® 209
   9.7   Summary.. 210
   References 210

Index 213

Preface

This book has its origin in the publication of the author's first book *Fuzzy Image Processing and Application with MATLAB*. In this book, an introduction to fuzzy set theory is included and different fuzzy techniques are discussed for image processing. Images are considered to be fuzzy for their grey levels or information lost while mapping and so fuzzy set theory is used, which considers uncertainty in the form of the membership function. But medical images contain a lot of uncertainties that are present in the edges/boundaries/regions. These images are poorly illuminated, and many structures are not clearly visible. So sets that use more uncertainties may be useful for medical image analysis.

Medical image analysis using advanced fuzzy set theoretic techniques is an exciting and dynamic branch of image processing. Since the introduction of fuzzy set theory by Prof. L.A. Zadeh in 1965, there has been an explosion of interest in advanced fuzzy set theories, such as intuitionistic fuzzy set and Type II fuzzy set, that represent the uncertainty in a better way. Fuzzy set theory considers only one uncertainty – the membership function. Prof. L.A. Zadeh proposed Type II fuzzy set theory that considers the fuzzy membership function as fuzzy. Later, Prof. K.T. Atanassov introduced intuitionistic fuzzy set theory that considers hesitation while defining the membership function. With intuitionistic fuzzy set, there has been a growth in image processing applications such as image enhancement, thresholding, clustering, edge detection and morphological image processing, but there is very little work on Type II fuzzy set.

There are many texts available in the market that deal with the fundamentals of image processing and also the use of fuzzy set theory in image processing. Bearing this in mind, an attempt has been made to write this book – *Medical Image Processing: Advanced Fuzzy Set Theoretic Techniques* – which deals with the application of intuitionistic fuzzy and Type II fuzzy set theories for medical image analysis.

The experimental results using intuitionistic fuzzy set theory and Type II fuzzy set theory are presented in the book. A concise summary of each chapter is presented, which highlights the ideas discussed in the chapter.

There is no such book on advanced fuzzy set theoretic techniques for medical image analysis. This book contains some preliminaries on intuitionistic fuzzy and interval Type II fuzzy set theories in two chapters. Then different methods are applied to enhance, segment/cluster or edge-detect medical images such as blood vessels, cells, tumours, clots and haemorrhages for image analysis.

This book is designed for graduate and doctoral-level students in universities worldwide. It is also extremely useful for teachers, engineers, scientists and all those who are interested in the field of medical image analysis. A preliminary knowledge on fuzzy set is required along with a good knowledge on mathematics and image processing. Nonetheless, sustained effort has been made to unify all concepts as they are introduced.

MATLAB® is a registered trademark of The MathWorks, Inc. For product information, please contact:

The MathWorks, Inc.
3 Apple Hill Drive
Natick, MA 01760-2098 USA
Tel: 508-647-7000
Fax: 508-647-7001
E-mail: info@mathworks.com
Web: www.mathworks.com

Organization of the Book

This book contains nine chapters. Each chapter begins with an introduction and a theory and then culminates with the application of advanced fuzzy techniques on medical images. To understand the techniques on intuitionistic and interval Type II fuzzy set theories, it is essential to have background knowledge on fuzzy sets and image processing. Chapter 1 introduces the fundamentals of intuitionistic fuzzy sets and Type II fuzzy sets. Essential operations and mathematics on intuitionistic fuzzy sets with a number of examples are explained. The fundamentals of interval Type II fuzzy sets are also explained. The chapter also includes intuitionistic fuzzy relations. Chapter 2 discusses, with examples, medical image processing where advanced fuzzy set theory is used. Also, medical image registration and image retrieval are detailed in the chapter. Chapter 3 covers fuzzy operators where fuzzy aggregation operators (weighted and ordered weighted averaging operator), t-norm, t-conorm, negation, intuitionistic fuzzy weighted averaging operator and hybrid operator are detailed. Chapter 4 covers different types of similarity, distance measures and entropy between intuitionistic fuzzy sets. Also, intuitionistic fuzzy divergence and intuitionistic fuzzy entropies are explained. Some information measures on interval Type II fuzzy sets are included. Intuitionistic fuzzy and interval Type II fuzzy enhancement is covered in detail in Chapter 5. In the enhancement method, structures that are not clearly visible in medical images are highlighted. It discusses the benefits of the use of advanced fuzzy sets. The most effective part is that different techniques, which are found in the literature on images using intuitionistic fuzzy/Type II fuzzy methods, are applied to medical images to gain a better knowledge about the effectiveness of these methods. The purpose of image enhancement is included where the images before and after enhancement are segmented to show that the segmented images after enhancement are better, helping doctors/physicians diagnose diseases accurately. MATLAB® examples are also included. Chapter 6 provides a detailed account of thresholding of medical images. It contains a summary of fuzzy thresholding techniques on medical images and a detailed description of intuitionistic fuzzy and Type II fuzzy thresholding techniques. Experimental results on different methods on pathological images for blood vessel/cell segmentation are also included. Moreover, leucocyte segmentation using intuitionistic fuzzy and Type II fuzzy methods that preserve the shape of the leucocytes is discussed, which is essential to identify the type of disease. MATLAB examples are also included. Chapter 7 deals with fuzzy clustering of medical images. Clustering is also done using kernel clustering, which groups similar pixels of the image in one group and different pixels in different groups. Kernel clustering is very robust. It clusters abnormal lesions almost accurately and even in the presence of noise. It transforms the input data to a high-dimensional feature space using a non-linear mapping function, and the kernel may be Gaussian or hypertangent or of any other type. Experimental results on fuzzy kernel clustering and intuitionistic fuzzy clustering are shown on medical images with and without noise. Chapter 8 introduces edge detection, where edges of the images are identified to preserve the structural information of the images. Different types of methods along with their results are discussed. Fuzzy edge generation using interval Type II fuzzy set is also discussed. MATLAB examples are also included. Chapter 9 deals with fuzzy mathematical morphology. Mathematical morphology already exists, but applications of fuzzy mathematical morphology on images are few. A detailed description on fuzzy mathematical morphology along with examples of medical images is presented. Some work on edge detection using fuzzy morphology is presented, but these techniques find it difficult to detect edges in medical images. New techniques are applied to dilate, erode or detect edges in medical images. MATLAB examples are also included.

All MATLAB examples mentioned above are available for download from the CRC Press Web site at http://www.crcpress.com/product/isbn/9781498700450

Acknowledgements

I gratefully acknowledge the Department of Biotechnology, Ministry of Science and Technology, Government of India, for their financial support. I also acknowledge Dr. Madhumala Ghosh, Indian Institute of Technology, Kharagpur, for providing me with many pathological images.

Finally, the book would not have been completed without the never-ending support of my parents and my beloved little daughter, Shruti De, who always wore a smiling face during my writing of this book, nor would it have been completed within the stipulated time without the support of my husband, Dr. Swades Kumar De.

Tamalika Chaira

Author

Dr. Tamalika Chaira received her bachelor's degree from Bihar Institute of Technology, Sindri, Jharkhand, India, and master's degree in electronics and communication engineering from Bengal Engineering and Science University, Shibpur, Howrah, India. She earned her PhD in image processing from the Department of Electrical and Electronics Engineering, Indian Institute of Technology, in 2003. She was a research associate in the signal and image processing group at the National Research Council (CNR), Pisa, Italy.

She is currently a research scientist in the Department of Biotechnology, Government of India, and the Indian Institute of Technology, Delhi, India. Her research interests include image segmentation, clustering, region extraction using fuzzy set, intuitionistic fuzzy set, Type II fuzzy set theory and medical imaging. She is an author of the book *Fuzzy Image Processing and Applications with MATLAB* published by CRC Press (Taylor & Francis Group). She has also published many papers related to fuzzy/intuitionistic fuzzy set/Type II fuzzy set on medical images in international journals and chapters in edited books published by Springer. She has received the prestigious National Award (Innovative Young Biotechnologist Award, 2010) from the Government of India. She is a member of Soft Computing in Image Processing (SCIP). Her biography is listed in *Marquis Who's Who in Science and Engineering* (2008), *Marquis Who's Who in the World* (2011–2014) and the International Biographical Centre, Cambridge, England (2008).

1

Intuitionistic Fuzzy Set and Type II Fuzzy Set

1.1 Introduction

Fuzzy set theory, proposed by Zadeh in 1965, takes into account membership degree and non-membership degree. Non-membership degree is the complement of membership degree. However, in real life, this linguistic negation does not satisfy the logical negation. The selection of membership degree depends on the user's choice, which may be Gaussian, triangular, exponential or any other. There is an uncertainty involved in defining the membership function. This is the reason different results are obtained with different membership functions. Subsequently, Atanassov suggested an advanced fuzzy set, which is an intuitionistic fuzzy set (IFS). In this set, non-membership degree is not equal to the complement of the membership degree; rather, it is less than or equal to the complement of the membership degree, due to the uncertainty in defining the membership degree.

Again, the membership function defined in this fuzzy set is not precise. Zadeh, in 1975, introduced another advanced fuzzy set called Type II fuzzy set that represents the uncertainty in a better way by considering the Type I (ordinary) fuzzy set 'fuzzy'.

Both IFS and Type II fuzzy set have been found useful in many real-time image applications such as medical images and remotely sensed images. These images are mostly poorly illuminated, where the regions/boundaries are vague or hardly visible, creating uncertainty. As IFS considers more than one (two) uncertainties – membership and non-membership degrees – and Type II fuzzy set considers the uncertainty in the membership function, these sets may be useful in processing such images.

1.2 Intuitionistic Fuzzy Set

A fuzzy set A in a finite set $X = \{x_1, x_2, ..., x_n\}$ may be represented mathematically as

$$A = \{(x, \mu_A(x)) \mid x \in X\}$$

where the function $\mu_A(x) : X \to [0, 1]$ is the measure of the degree of belongingness or the membership function of an element x in the finite set X, and the measure of non-belongingness is $1 - \mu_A(x)$.

An IFS A in a finite set X may be mathematically represented as

$$A = \{(x, \mu_A(x), \nu_A(x)) \mid x \in X\} \tag{1.1}$$

where the functions $\mu_A(x)$, $\nu_A(x) : X \to [0, 1]$ are, respectively, the membership function and the non-membership function of an element x in a finite set X with the necessary condition

$$0 \le \mu_A(x) + \nu_A(x) \le 1$$

It is clear that every fuzzy set is a particular case of IFS:

$$A = \{(x, \mu_A(x), 1 - \mu_A(x)) \mid x \in X\}$$

Atanassov [1] also stressed the necessity of taking into consideration a third parameter $\pi_A(x)$, known as the intuitionistic fuzzy index or hesitation degree, which arises due to the lack of knowledge or 'personal error' in assigning the membership degree. Therefore, with the introduction of a hesitation degree, an IFS A in X may be represented as

$$A = \{(x, \mu_A(x), \nu_A(x), \pi_A(x)) \mid x \in X\}$$

with the condition

$$\pi_A(x) + \mu_A(x) + \nu_A(x) = 1 \tag{1.2}$$

It is obvious that

$$0 \le \pi_A(x) \le 1, \quad \text{for each } x \in X$$

The following definitions hold for all IFSs A and B of set X:

1. $A \cup B = \{\max(\mu_A, \mu_B), \min(\nu_A, \nu_B)\}$
2. $A \cap B = \{\min(\mu_A, \mu_B), \max(\nu_A, \nu_B)\}$
3. $A \prec B = \{x, \mu_A(x) < \mu_B(x), \nu_A(x) < \nu_B(x)\}$
4. $\bar{A} = \{x, \nu_A(x), \mu_A(x)\}$
5. $A \leq B = \{x, \mu_A(x) \leq \mu_B(x), \nu_A(x) \geq \nu_B(x)\}$
6. $A \cdot B = \{x, \mu_A(x) \cdot \mu_B(x), \nu_A(x) + \nu_B(x) - \nu_A(x) \cdot \nu_B(x)\}$

1.3 Some Operations on Intuitionistic Fuzzy Sets

Similar to fuzzy sets, IFSs have also undergone some operations. Operations on IFS have been carried out by many authors [8,16,18]. The negation 'NOT'; the connectives 'AND' and 'OR'; hedges 'VERY', 'MORE OR LESS', 'VERY VERY' and 'VERY HIGHLY'; and other terms are used to represent linguistic variables. For two IFSs A and B, with $\mu(x)$ and $v(x)$ as the membership and non-membership degrees of the elements in two sets, the following conditions hold:

$$A \text{ and } B = A \wedge B = \{x, \min(\mu_A(x), \mu_B(x)), \max(\nu_A(x), \nu_B(x))\}$$

$$A \text{ or } B = A \vee B = \{x, \max(\mu_A(x), \mu_B(x)), \min(\nu_A(x), \nu_B(x))\}$$

The product of the two IFSs A and B is given by

$$A \cdot B = \{(x, \mu_A(x) \cdot \mu_B(x), \nu_A(x) + \nu_B(x) - \nu_A(x) \cdot \nu_B(x))\}$$

$$A^2 = A \cdot A = \{x, [\mu_A(x)]^2, 1 - (1 - \nu_A(x))^2\}$$

It is also called concentration of set A, CON(A):

$$A^{1/2} = A \cdot A = \{x, [\mu_A(x)]^{1/2}, 1 - (1 - \nu_A(x))^{1/2}\}$$

or the dilation of set A, that is, DIL(A):

$$A^3 = \{x, [\mu_A(x)]^3, 1 - (1 - \nu_A(x))^3\}$$

Therefore, in general for any positive integer n,

$$A^n = \{x, [\mu_A(x)]^n, 1-(1-\nu_A(x))^n\}$$

and follows the condition

$$0 \le [\mu_A(x)]^n + [1-(1-\nu_A(x))^n] \le 1$$

Like fuzzy sets, for IFSs, linguistic hedges can be defined:

CON(A) = very (A)
DIL(A) = more or less A
plus(A) = $A^{1.25}$
minus(A) = $A^{0.75}$

Example 1.1

Let A be an IFS denoting the linguistic variable 'YOUNG' in the universe of discourse [0, 100]:

$$A(\text{YOUNG}) = \{(x, \mu_{\text{YOUNG}(A)}(x), \nu_{\text{YOUNG}(A)}(x)) \mid x \in U\}$$

where

$$\mu_{\text{YOUNG}}(x) = 1, \qquad\qquad\qquad x \in [0, 28]$$

$$= \frac{1}{\left(1+((x-28)/5)^2\right)}, \quad x \in [28, 100]$$

$$\nu_{\text{YOUNG}}(x) = 0, \qquad\qquad\qquad x \in [0, 30]$$

$$= 1 - \frac{1}{\left(1+((x-30)/5)^2\right)}, \quad x \in [30, 100]$$

'VERY YOUNG' may be denoted as

$$\text{VERY YOUNG} = \{(x, \mu_{\text{VERY YOUNG}}(x), \nu_{\text{VERY YOUNG}}(x)) \mid x \in U\}$$

where

$$\mu_{\text{VERY YOUNG}}(x) = 1, \qquad\qquad\qquad x \in [0, 28]$$

$$= \frac{1}{\left(1+((x-28)/5)^2\right)^2}, \quad x \in [28, 100]$$

$$\nu_{\text{VERY YOUNG}}(x) = 0, \qquad\qquad\qquad x \in [0, 30]$$

$$= 1 - \frac{1}{\left(1 + ((x-30)/5)^2\right)^2}, \quad x \in [30, 100]$$

Likewise, 'MORE OR LESS YOUNG' may be defined as

$$\mu_{\text{MORE OR LESS YOUNG}}(x) = 0, \qquad\qquad x \in [0, 28]$$

$$= \frac{1}{\left(1 + ((x-28)/5)^2\right)^{1/2}}, \quad x \in [28, 100]$$

$$\nu_{\text{MORE OR LESS YOUNG}}(x) = 1, \qquad\qquad x \in [0, 30]$$

$$= 1 - \frac{1}{\left(1 + ((x-30)/5)^2\right)^{1/2}}, \quad x \in [30, 100]$$

1.4 Fuzzy Complement and Intuitionistic Fuzzy Generator

Fuzzy complement is used in creating IFSs. Fuzzy complement is a fuzzy negation. Suppose $\mu(x)$ is defined as the degree to which x belongs to A. Let $c \cdot \mu(x)$ denote a fuzzy complement of A that signifies the degree to which x belongs to the fuzzy complement set cA. Then the complement cA is defined by a function $c: [0, 1] \rightarrow [0, 1]$.

The fuzzy complement operator has the following properties:

Boundary condition

$$c(0) = 1 \quad \text{and} \quad c(1) = 0$$

Involutive property

$$c \cdot c(\mu(x)) = \mu(x)$$

Monotonicity

For all $a, b \in [0, 1]$, if $a \leq b$, then $c(a) \geq c(b)$

$c \cdot \mu(x)$ is continuous

The fuzzy complement function has been studied by many authors [8,15,17].

The practical fuzzy systems use the standard Zadeh's fuzzy negation as

$$c \cdot \mu(x) = 1 - \mu(x) \tag{1.3}$$

This is the simplest and is extensively used in fuzzy set theory.

Fuzzy complement is computed from the fuzzy complement functional, which is defined as

$$N(\mu(x)) = g^{-1}(g(1) - g(\mu(x))) \tag{1.4}$$

where $g(\cdot)$ is an increasing function with $g(0) = 0$.

Sugeno-type fuzzy complement [15] is generated using the same fuzzy complement function with an increasing function, which is given as

$$g(\mu(x)) = \frac{1}{\lambda} \log(1 + \lambda \mu(x)) \tag{1.5}$$

From Equation 1.4,

$$N(\mu(X)) = g^{-1} \left[\frac{1}{\lambda} \log(1 + \lambda) - \frac{1}{\lambda} \log(1 + \lambda \mu(x)) \right]$$

or

$$N(\mu(x)) = g^{-1} \left(\frac{1}{\lambda} \log \frac{1 + \lambda}{1 + \lambda \mu(x)} \right)$$

From Equation 1.5,

$$g^{-1}(\mu(x)) = \frac{e^{\lambda \mu(x)} - 1}{\lambda}$$

Then by induction method,

$$g^{-1} \left(\frac{1}{\lambda} \log \frac{1 + \lambda}{1 + \lambda \cdot \mu(x)} \right) = \frac{e^{\lambda \cdot \left(\frac{1}{\lambda} \log \frac{1 + \lambda}{1 + \lambda \cdot \mu(x)} \right)} - 1}{\lambda}$$

$$= \frac{\dfrac{1 + \lambda}{1 + \lambda \cdot \mu(x)} - 1}{\lambda} = \frac{1 - \mu(x)}{1 + \lambda \cdot \mu(x)}$$

So,

$$N(\mu(x)) = \frac{1-\mu(x)}{1+\lambda \cdot \mu(x)}, \quad \lambda \in (-1, \infty) \tag{1.6}$$

Likewise, Yager's [17] class can also be generated using this fuzzy complement function using an increasing function
$g(\mu(x)) = \mu(x)^\alpha$ and the fuzzy complement is

$$N(\mu(x)) = [1-\mu(x)^\alpha]^{1/\alpha}, \quad \alpha \in (0, \infty) \tag{1.7}$$

Equations 1.6 and 1.7 reduce to Equation 1.3, when $\alpha = 1$ and $\lambda = 0$.
Another form of increasing function is

$$g(\mu(x)) = \frac{\mu(x)}{\gamma + (1-\gamma)\mu(x)}$$

and its complement function is

$$N(\mu(x)) = \frac{\gamma^2(1-\mu(x))}{\gamma^2(1-\mu(x)) + \mu(x)}, \quad \gamma > 0 \tag{1.8}$$

At $\gamma = 1$, the fuzzy complement in Equation 1.8 reduces to the standard fuzzy complement.
Roychowdhury and Pedrycz [14] suggested a different type of fuzzy complement function:

$$N(\mu(x)) = g^{-1}\left(\frac{-g(\mu(x))}{1+g(\mu(x))}\right)$$

where $g: [0, 1] \rightarrow (-\infty, -1)$ is a continuous function with $g(0) = -\infty$ and $g(1) = -1$ if it is strictly increasing, and $g(0) = 1$ and $g(1) = -\infty$ if it is strictly decreasing.
Klir and Yuan [10] suggested a dual generator:

$$N(\mu(x)) = f^{-1}(f(0) - f(\mu(x)))$$

where $f(\cdot)$ is a decreasing function.
Example: For Yager's class of fuzzy complement, the decreasing generating function is

$$f(\mu(x)) = 1 - \mu(x)^\omega$$

$$\text{So, } f^{-1}(\mu(x)) = (1-\mu(x))^{1/\omega}$$

$$N(\mu(x)) = f^{-1}(1-1+(\mu(x))^{\omega}) = f^{-1}((\mu(x))^{\omega})$$

$$\text{So, } N(\mu(x)) = f^{-1}((\mu(x))^{\omega}) = \left(1-(\mu(x))^{\omega}\right)^{1/\omega}$$

Likewise, for Sugeno's class, the decreasing function is

$$f(\mu(x)) = \ln(1+\mu(x)) - \frac{1}{\lambda}\log(1+\lambda\mu(x)), \quad \lambda > -1$$

For the standard fuzzy complement, the decreasing generating function is $f(\mu(x)) = -k \cdot \mu(x) + k, k > 0$.

1.5 Intuitionistic Fuzzy Generator

Not all fuzzy complements are intuitionistic fuzzy generator. In order to construct Atanassov's IFS from fuzzy set theory, intuitionistic fuzzy generators are used. From the definition of intuitionistic fuzzy generator given by Bustince and Burillo [5], a function φ: [0, 1] will be called an intuitionistic fuzzy generator if

$$\varphi(x) \leq 1 - x, \quad \forall x \in [0, 1] \tag{1.9}$$

So, according to the definition, φ(0) ≤ 1 and φ(1) = 0.

In an IFS, two uncertainties – membership and non-membership degrees – are considered and the non-membership degree is not a complement of the membership degree. This is due to the uncertainty present in the membership function. The less than or equal to sign in Equation 1.9 is due to the hesitation degree.

Non-membership values may be calculated from Sugeno, Yager, or Chaira, or any other type of intuitionistic fuzzy generator. If we take the example of Sugeno's fuzzy complement,

$$N(\mu(x)) = \varphi_\lambda(\mu(x)) = \frac{1-\mu(x)}{1+\lambda \cdot \mu(x)}$$

At −1 < λ < 0, the condition for an intuitionistic fuzzy generator does not hold.

Also for Yager's fuzzy complement,

$$N(\mu(x)) = \varphi_\lambda(\mu(x)) = \left[1-\mu(x)^\alpha\right]^{1/\alpha}$$

At $\alpha > 1$, the condition for intuitionistic fuzzy generator does not hold. At these conditions, it follows that $\mu(x) + N(\mu(x)) > 1$, which is not true. For intuitionistic fuzzy generator, the condition of λ, α is changed. For Yager-type intuitionistic fuzzy generator, the condition is

$$\varphi(\mu(x)) = [1-\mu(x)^\alpha]^{1/\alpha}, \quad 0<\alpha<1$$

and for Sugeno type,

$$\varphi(\mu(x)) = \frac{1-\mu(x)}{1+\lambda\cdot\mu(x)}, \quad \lambda\geq 0$$

Thus, with the help of the Sugeno-type intuitionistic fuzzy complement, IFS becomes

$$A_\lambda^{IFS} = \left\{x, \mu_A(x), \frac{(1-\mu_A(x))}{(1+\lambda\cdot\mu_A(x))}\bigg| x \in X\right\}$$

with hesitation degree

$$\pi_A(x) = 1-\mu_A(x)-\frac{(1-\mu_A(x))}{(1+\lambda\cdot\mu_A(x))}$$

Since the denominator, $1 + \lambda\cdot\mu_A(x)$, in the non-membership term $(1 - \mu_A(x))/(1 + \lambda\cdot\mu_A(x))$ is greater than 1, the non-membership term is less than $1 - \mu(x)$ for all $x \in X$.

Likewise, with Yager's intuitionistic fuzzy generator, IFS becomes

$$A_\lambda^{IFS} = \{x, \mu_A(x), (1-\mu_A(x)^\alpha)^{1/\alpha} \mid x \in X\}$$

Chaira [7] also suggested an intuitionistic fuzzy generator as follows:

$$N(\mu(x)) = \frac{1-\mu(x)}{1-(1-e^\lambda)\mu(x)} = \frac{1-\mu(x)}{1+(e^\lambda-1)\mu(x)}, \quad \lambda > 0 \qquad (1.10)$$

with an increasing generating function as

$$f(\mu(x)) = \frac{1}{\lambda}\ln[1-\mu(x)(1-e^{\lambda})], \quad \lambda > 0 \quad \text{and} \quad \mu(x) \in [0,1]$$

With Chaira's intuitionistic fuzzy generator, IFS may be written as

$$A_{\lambda}^{\text{IFS}} = \left\{ x, \mu_A(x), \frac{1-\mu(x)}{1+(e^{\lambda}-1)\mu(x)} \middle| x \in X \right\}$$

1.6 Intuitionistic Fuzzy Relations

Since 1965, when Zadeh introduced fuzzy set theory, researchers modelled fuzzy relation (FR) on X, that is, a function $R: X \times X \rightarrow [0, 1]$, and used FR in different fields. In 1983, when Atanassov introduced IFS, many researchers extended FR using IFS and modelled intuitionistic fuzzy relations (IFRs) on X. Burillo and Bustince [2–4] had given the definition of IFRs and their properties. Lei et al. [11] further explored IFRs and its compositional operations. To fulfil the properties of IFRs, t-norm and t-conorm are required.

Let X and Y be two universes of discourse and IFS($X \times Y$) represent the family of all IFSs in $X \times Y$. Let $R \in$ IFS($X \times Y$) be the IFR, which is a subset in $X \times Y$, given as

$$R = \{(x,y), \mu_R(x,y), \nu_R(x,y) \mid x \in X, y \in Y\}$$

where

$$\mu_R : X \times Y \rightarrow [0,1]$$

$$\nu_R : X \times Y \rightarrow [0,1]$$

denote the membership and non-membership functions of R, respectively, that satisfy the condition $0 \le \mu_R(x, y) + \nu_R(x, y) \le 1$ and $\pi_R(x, y) = 1 - \mu_R(x, y) - \nu_R(x, y)$ is the hesitation index.

The complementary relation of R is

$$R_c = \{(x,y), \nu_R(x,y), \mu_R(x,y) \mid (x,y) \in X \times Y\}$$

The following operations on IFRs are as follows:

1. If R is an IFR on $X \times Y$, then R^{-1} is an IFR on $Y \times X$:

$$\mu_{R^{-1}}(y,x) = \mu_R(x,y) \quad \text{and} \quad \nu_{R^{-1}}(y,x) = \nu_R(x,y), \quad \forall(y,x) \in Y \times X$$

This is also called inverse relation on R.

2. Union

$$\mu_{R\vee Q} = \max(\mu_R(x,y), \mu_Q(x,y))$$

$$\nu_{R\vee Q} = \min(\nu_R(x,y), \nu_Q(x,y))$$

3. Intersection

$$\mu_{R\wedge Q} = \min(\mu_R(x,y), \mu_Q(x,y))$$

$$\nu_{R\wedge Q} = \max(\nu_R(x,y), \nu_Q(x,y))$$

4. For three elements, R, P, and Q, on relation IFR $\in (X \times Y)$

$$(R \vee P)^{-1} = R^{-1} \vee P^{-1}$$

$$(R \wedge P)^{-1} = R^{-1} \wedge P^{-1}$$

$$R \vee (P \wedge Q) = (R \vee P) \wedge (R \vee Q)$$

1.7 Composition of Intuitionistic Fuzzy Relation (Supremum–Infimum)

An IFR R from X to Y is an IFS of $X \times Y$ characterised by a membership function μ_R and a non-membership function ν_R. An IFR R from X to Y will be denoted by $R(X \rightarrow Y)$.

Let $P(X \rightarrow Y)$ and $R(Y \rightarrow Z)$ be two IFRs. The supremum–infimum composition $R \circ P$ is an IFR of X to Z. It is defined in terms of the membership and non-membership degrees as

$$\mu_{R\circ P}(x,z) = \underset{y}{\vee}\{\mu_P(x,y) \wedge \mu_R(y,z)\}$$

$$\nu_{R\circ P}(x,z) = \underset{y}{\wedge}\{\nu_P(x,y) \vee \nu_R(y,z)\}$$

(1.11)

respectively, where $\wedge$ denotes infimum and $\vee$ denotes supremum.

1.8 Composition of Intuitionistic Fuzzy Relation Using Fuzzy *t*-Norm and *t*-Conorm

Composition of IFR can also be carried out using *t*-norms and *t*-conorms [2–4]. *T*-norms and *t*-conorms are a kind of binary operation used to represent the intersection and union in fuzzy set theory, respectively. In this section, fuzzy *t*-norm and *t*-conorm are discussed briefly, and these are discussed in detail in Chapter 3.

1.8.1 *t*-Norm

A *t*-norm T: $[0, 1] \rightarrow [0, 1]$ is a kind of binary operation used in the framework of fuzzy logic and probabilistic metric spaces. It represents the intersection in fuzzy set theory or an 'ANDing' operator.

The four basic *t*-norms are

1. The minimum: $T(x, y) = \min(x, y)$
2. The product: $T(x, y) = x \cdot y$
3. The Lukasiewicz *t*-norm: $T_L(x, y) = \max(x + y - 1, 0)$
4. The Nilpotent minimum *t*-norm:

$$T(x,y) = \min(x,y) \quad \text{if } x + y > 1$$

$$= 0 \qquad\qquad \text{otherwise}$$

1.8.2 *t*-Conorm

A *t*-conorm T: $[0, 1] \rightarrow [0, 1]$ is a kind of binary operation used in the framework of fuzzy logic and probabilistic metric spaces. It represents union in fuzzy set theory or an 'ORing' operator. The four basic *t*-conorms are

1. The maximum: $S(x, y) = \max(x, y)$
2. The product: $S(x, y) = x + y - x \cdot y$
3. The Lukasiewicz *t*-conorm: $T_L(x, y) = \min(x + y, 1)$
4. Nilpotent minimum *t*-conorm:

$$S(x,y) = \max(x,y) \quad \text{if } x + y < 1$$

$$= 1 \qquad\qquad \text{otherwise}$$

For IFRs, *t*-norms and *t*-conorms will be designated with letters α, β, ρ and δ in this chapter.

Let α, β, ρ and δ be t-norms or t-conorms and the relations $R \in \text{IFR}(X \times Y)$ and $P \in \text{IFR}(Y \times Z)$. The composed relation $R \overset{\alpha,\beta}{\underset{\rho,\delta}{\circ}} P \in \text{IFR}(X \times Z)$ is written as [2]

1. $R \overset{\alpha,\beta}{\underset{\rho,\delta}{\circ}} P = \left\{ (x,z), \mu_{R \overset{\alpha,\beta}{\underset{\rho,\delta}{\circ}} P}(x,z), v_{R \overset{\alpha,\beta}{\underset{\rho,\delta}{\circ}} P}(x,z) \middle| x \in X, z \in Z \right\}$ (1.12)

where

$$\mu_{R \overset{\alpha,\beta}{\underset{\rho,\delta}{\circ}} P}(x,z) = \alpha\{\beta[\mu_R(x,y), \mu_P(y,z)]\}$$

$$v_{R \overset{\alpha,\beta}{\underset{\rho,\delta}{\circ}} P}(x,z) = \rho\{\delta[\mu_R(x,y), \mu_P(y,z)]\}$$

(1.13)

and

$$0 \le \mu_{R \overset{\alpha,\beta}{\underset{\rho,\delta}{\circ}} P}(x,z) + v_{R \overset{\alpha,\beta}{\underset{\rho,\delta}{\circ}} P}(x,z) \le 1, \quad \forall(x,z) \in X \times Z$$

condition holds.

The choice of t-norm and t-conorm should be such that the earlier condition, that is,

$$0 \le \mu_{R \overset{\alpha,\beta}{\underset{\rho,\delta}{\circ}} P}(x,z) + v_{R \overset{\alpha,\beta}{\underset{\rho,\delta}{\circ}} P}(x,z) \le 1, \quad \forall(x,z) \in X \times Z$$

holds.

It is to be noted that α and β are applied for membership functions and ρ and δ are applied to non-membership functions. But the composition of IFR satisfies most of the properties for $\alpha = \vee$, β t-norm, $\rho = \wedge$, δ t-conorm.

2. If $\alpha = \vee$, $\beta = \wedge$ and $\rho = \wedge$, $\delta = \vee$, then Equation 1.13 reduces to Equation 1.11:

$$\mu_{R \circ P}(x,z) = \underset{y}{\vee}\{\mu_R(x,y) \wedge \mu_P(y,z)\}$$

$$v_{R \circ P}(x,z) = \underset{y}{\wedge}\{v_R(x,y) \vee v_P(y,z)\}$$

3. If R, $P \in \text{IFR}(Y \times Z)$ and $Q \in \text{IFR}(X \times Y)$ and α, β, ρ and δ are the t-norms or t-conorms, the following properties hold:

$$(R \vee P) \overset{\alpha,\beta}{\underset{\rho,\delta}{\circ}} Q \ge \left(R \overset{\alpha,\beta}{\underset{\rho,\delta}{\circ}} Q \right) \vee \left(P \overset{\alpha,\beta}{\underset{\rho,\delta}{\circ}} Q \right)$$

$$(R \wedge P) \overset{\alpha,\beta}{\underset{\rho,\delta}{\circ}} Q \leq \left(R \overset{\alpha,\beta}{\underset{\rho,\delta}{\circ}} Q \right) \wedge \left(P \overset{\alpha,\beta}{\underset{\rho,\delta}{\circ}} Q \right)$$

4. Again, for each $R \in$ IFR$(X \times Y)$ and $P \in$ IFR$(Y \times Z)$ and α, β, ρ and δ as any *t*-norms or *t*-conorms, then

$$\left(P \overset{\alpha,\beta}{\underset{\rho,\delta}{\circ}} P \right)^{-1} = R^{-1} \overset{\alpha,\beta}{\underset{\rho,\delta}{\circ}} P^{-1} \tag{1.14}$$

1.8.3 Reflexive Property

IFR R is said to be reflexive

$$\text{iff } \forall x \in X, \mu_R(x,x) = 1 \quad \text{and} \quad v_R(x,x) = 0$$

and antireflexive if

$$\forall x \in X, v_R(x,x) = 1 \quad \text{and} \quad \mu_R(x,x) = 0 \tag{1.15}$$

For α *t*-conorm, β *t*-norm, ρ *t*-norm and δ *t*-conorms

If $R \in$ IFR$(X \times X)$ is reflexive, then $R \leq R \overset{\alpha,\beta}{\underset{\rho,\delta}{\circ}} R$

If $R \in$ IFR$(X \times X)$ is antireflexive, then $R \geq R \overset{\rho,\delta}{\underset{\alpha,\beta}{\circ}} R$

Example 1.2

The following example shows that the relation R is not reflexive but follows the reflexive property:

$$\mu_R = \begin{pmatrix} 0.4 & 0.7 & 0.2 \\ 0.5 & 0.9 & 0.5 \\ 0.1 & 0.4 & 0.1 \end{pmatrix}, \quad v_R = \begin{pmatrix} 0.5 & 0.2 & 0.8 \\ 0.2 & 0.0 & 0.4 \\ 0.6 & 0.3 & 0.8 \end{pmatrix}$$

For $\alpha = \vee$, $\beta = \wedge$, $\rho = \wedge$ and $\delta = \vee$, we have

$$\mu_{\substack{\vee,\wedge \\ R \circ R \\ \wedge,\vee}} = \begin{pmatrix} 0.5 & 0.7 & 0.5 \\ 0.5 & 0.9 & 0.5 \\ 0.4 & 0.4 & 0.4 \end{pmatrix}, \quad v_{\substack{\vee,\wedge \\ R \circ R \\ \wedge,\vee}} = \begin{pmatrix} 0.2 & 0.2 & 0.4 \\ 0.2 & 0.0 & 0.4 \\ 0.2 & 0.2 & 0.4 \end{pmatrix}$$

It satisfies the reflexive property as $\mu_{\substack{R \circ R \\ \wedge,\vee}}^{\vee,\wedge}(x,z) \geq \mu_R(x,z)$ and $v_{\substack{R \circ R \\ \wedge,\vee}}^{\vee,\wedge}(x,z) \leq v_R(x,z)$, but R is not reflexive as $\mu_R(x,x) \neq 1$.

1.8.4 Symmetric Property

IFR R is said to be symmetric iff $R = R^{-1}$, that is,

$$\forall x,y \in X, \quad \mu_R(x,y) = \mu_R(y,x)$$
$$v_R(x,y) = v_R(y,x)$$

Relation R is π-symmetric or antisymmetric if

$$\forall (x,y) \in X \times X, x \neq y, \quad \text{then } \mu_R(x,y) \neq \mu_R(y,x)$$
$$v_R(x,y) \neq v_R(y,x)$$
$$\pi_R(x,y) = \pi_R(y,x)$$

If α, β, ρ and δ are any t-norms or t-conorms and R, $P \in$ IFR($X \times X$) is symmetrical, then

$$R \underset{\rho,\delta}{\overset{\alpha,\beta}{\circ}} P = \left(P \underset{\rho,\delta}{\overset{\alpha,\beta}{\circ}} R \right)^{-1}$$

Also, if R is symmetrical, $R \underset{\rho,\delta}{\overset{\alpha,\beta}{\circ}} R$ is also symmetrical. But the composition of two symmetrical relations will not always be symmetrical.

1.8.5 Transitive Property

IFR R is said to be transitive if $R \circ R \subseteq R$ or $R^2 \subseteq R$ where $\subseteq$ denotes

$$A \subseteq B = \{x, \mu_A(x) \leq \mu_B(x), v_A(x) \geq v_B(x)\}$$

For α t-conorm, β t-norm, ρ t-norm, and δ t-conorm,

Relation $R \in$ IFR($X \times X$) is transitive if $R \geq R \underset{\rho,\delta}{\overset{\alpha,\beta}{\circ}} R$ or $R \geq R \underset{\wedge,\delta}{\overset{\vee,\beta}{\circ}} R$

Relation R is c-transitive if $R \leq R \underset{\alpha,\beta}{\overset{\rho,\delta}{\circ}} R$ or $R \leq R \underset{\vee,\beta}{\overset{\wedge,\delta}{\circ}} R$

Transitive closure of R is the minimum IFR $\hat{R}$ on $X \times X$ which contains R and it is transitive, that is, $R \leq \hat{R}$ and $\hat{R} \underset{\rho,\delta}{\overset{\alpha,\beta}{\circ}} \hat{R} \leq \hat{R}$.

c-Transitive closure of the relation R is the biggest c-transitive relation $\check{R}$ on $X \times X$ which contains R and is transitive.

Transitive closure $\hat{R}$ of IFR R is defined as

$$\hat{R} = R \cup R^2 \cup R^3 \cup \cdots \tag{1.16}$$

For every $R \in \text{IFR}(X \times X)$, it follows that if $\alpha = \vee$, $\beta = \wedge$, $\rho = \wedge$, and $\delta = \vee$, then

1. $\hat{R} = R \vee R \overset{\vee,\wedge}{\underset{\wedge,\vee}{\circ}} R \vee R \overset{\vee,\wedge}{\underset{\wedge,\vee}{\circ}} R \overset{\vee,\wedge}{\underset{\wedge,\vee}{\circ}} R \vee \cdots \vee R^n$

2. $\check{R} = R \vee R \overset{\wedge,\vee}{\underset{\vee,\wedge}{\circ}} R \vee R \overset{\wedge,\vee}{\underset{\vee,\wedge}{\circ}} R \overset{\wedge,\vee}{\underset{\vee,\wedge}{\circ}} R \vee \cdots \vee R^n$

1.9 Interval-Valued Intuitionistic Fuzzy Set

Sometimes, it may happen that the membership degrees are not exactly defined but a vague range is defined. An interval-valued intuitionistic fuzzy set (IVIFS) is defined as

$$A = \{(x, \mu_A(x), v_A(x)) \,|\, x \in X\}$$

where
$$\mu_A(x) : X \to (0, 1)$$

$$v_A(x) : X \to (0, 1)$$

and the interval $(0, 1)$ denotes the closed subinterval in the interval $[0, 1]$. This implies that the membership and non-membership degrees lie in an interval range with the condition

$$0 \le \sup(\mu_A(x)) + \sup(v_A(x)) \le 1$$

Let $\mu_A(x) = [\mu_A^-(x), \mu_A^+(x)]$ and $v_A(x) = [v_A^-(x), v_A^+(x)]$, then

$$A^{\text{IVIFS}} = \left\{ x, \left[\mu_A^-(x), \mu_A^+(x)\right], \left[v_A^-(x), v_A^+(x)\right] \middle| x \in X \right\}$$

with $0 \le \mu_A^+(x) + v_A^+(x) \le 1$.

The hesitation degree also lies in an interval which is given as

$$\left[\pi_A^-(x), \pi_A^+(x)\right] = \left[1 - \mu_A^-(x) - v_A^-(x), 1 - \mu_A^+(x) - v_A^+(x)\right]$$

If $\mu_A^-(x) = \mu_A^+(x)$ and $v_A^-(x) = v_A^+(x)$, then IVIFS becomes IFS.
For two interval-valued fuzzy sets,

$$A^{\text{IVIFS}} = \left\{ x, \left[\mu_A^-(x), \mu_A^+(x)\right], \left[v_A^-(x), v_A^+(x)\right] \middle| x \in X \right\}$$

and

$$B^{\text{IVIFS}} = \left\{ x, \left[\mu_B^-(x), \mu_B^+(x)\right], \left[v_B^-(x), v_B^+(x)\right] \middle| x \in X \right\}$$

the following relation holds:

1. $A \subseteq B$, iff $\mu_A^-(x) \leq \mu_B^-(x), \mu_A^+(x) \leq \mu_B^+(x), v_A^-(x) \geq v_B^-(x), v_A^+(x) \geq v_B^+(x)$

   for $x \in X$

2. $A = B$, iff $A \subseteq B$ and $B \subseteq A$

3. $A_{\text{IVIFS}}^c = \{x, [v_A^-(x), v_A^+(x)], [\mu_A^-(x), \mu_A^+(x)] \mid x \in X\}$

4. $A \cap B = [\min(\mu_A^-(x), \mu_B^-(x)), \min(\mu_A^+(x), \mu_B^+(x))],$

   $[\max(v_A^-(x), v_B^-(x)), \max(v_A^+(x), v_B^+(x))]$

5. $A \cup B = [\max(\mu_A^-(x), \mu_B^-(x)), \max(\mu_A^+(x), \mu_B^+(x))],$

   $[\min(v_A^-(x), v_B^-(x)), \min(v_A^+(x), v_B^+(x))]$

6. $A + B = [(\mu_A^-(x) + \mu_B^-(x) - \mu_A^-(x) \cdot \mu_B^-(x)), (\mu_A^+(x) + \mu_B^+(x) - \mu_A^+(x) \cdot \mu_B^+(x))]$

   $[v_A^-(x) \cdot v_B^-(x), v_A^+(x) \cdot v_B^+(x)]$

7. $A^\lambda = [\mu_A^{-\lambda}(x), \mu_A^{+\lambda}(x), 1 - (1 - v_A^-(x))^\lambda, 1 - (1 - v_A^+(x))^\lambda]$

8. $A \cdot B = [\mu_A^-(x) \cdot \mu_B^-(x), \mu_A^+(x) \cdot \mu_B^+(x)],$

   $[v_A^-(x) + v_B^-(x) - v_A^-(x) \cdot v_B^-(x), v_A^+(x) + v_B^+(x) - v_A^+(x) \cdot v_B^+(x)]$

Correlation coefficient using interval valued intuitionistic fuzzy sets (IVIFS) is studied by authors [6,13]. For two IVIFS A and B, if

$$C(A, B) = \frac{1}{2} \sum \left[\mu_A^-(x) \cdot \mu_B^-(x) + \mu_A^+(x) \cdot \mu_B^+(x) + v_A^-(x) \cdot v_B^-(x) + v_A^+(x) \cdot v_B^+(x)\right], \text{ then}$$

$$\text{Corr}(A, B) = \frac{C(A, B)}{\sqrt{C(A, A) \cdot C(B \cdot B)}}$$

1.10 Type II Fuzzy Set

Membership functions are usually defined by the expert and are based on his or her intuition or knowledge. So, different fuzzy techniques differ mainly in the way that they define the membership function, and so different results are obtained using different types of membership functions. To find a more robust solution, Type II fuzzy sets are introduced. There are different sources of uncertainties in type I fuzzy sets; they are (1) inaccurate measurements, (2) disagreement of the membership values with the accurate membership values of the data that are used to tune Type II fuzzy set and (3) uncertainty in the location, shape or other parameters. According to Mendel and John [12], 'Type I fuzzy sets are not able to directly model such uncertainties because their membership functions are totally crisp. On the other hand, Type II fuzzy sets are able to model such uncertainties because their membership functions are themselves fuzzy'. If there is no uncertainty, then a Type II fuzzy set reduces to a type I fuzzy set. Type II fuzzy set is obtained by blurring the type I fuzzy set. So, Type II fuzzy sets are the fuzzy sets for which the membership function is not a single value for every element but an interval. A Type II fuzzy set may be written as

$$A_{\text{TypeII}} = \{x, \widehat{\mu}_A(x) \mid x \in X\}$$

where $\widehat{\mu}_A(x)$ is the Type II membership function. The footprint of uncertainty (FOU) represents the uncertainty in the primary memberships of the Type II fuzzy set as shown in Figure 1.1 [9].

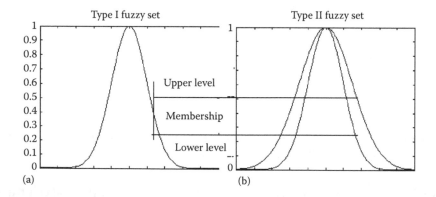

FIGURE 1.1
(a) Type I membership function and (b) interval of Type II fuzzy membership function in the FOU.

An interval-based Type II fuzzy set is constructed by defining upper and lower membership values of type I fuzzy set. So, the membership function is in an interval range, that is, fuzzy, and is written as

$$\mu^{\text{upper}} = [\mu(x)]^\alpha$$
$$\mu^{\text{lower}} = [\mu(x)]^{1/\alpha}$$

(1.17)

where $0 < \alpha \le 1$.

The points within the membership areas that is in the interval region are represented by type I fuzzy set. So, a more practical form of representing Type II fuzzy set is

$$A_{\text{TypeII}} = \{x, \mu_U(x), \mu_L(x) \mid x \in X\}$$

and

$$\mu_U(x) < \mu(x) < \mu_L(x), \quad \mu \in [0, 1]$$

where upper and lower limits may be the linguistic hedges like dilation or concentration:

$$\mu_U(x) = \mu(x)^{0.5}$$
$$\mu_L(x) = \mu(x)^2$$

or it can be deaccentuation or accentuation:

$$\mu_U(x) = \mu(x)^{0.75}$$
$$\mu_L(x) = \mu(x)^{1.25}$$

As the linguistic hedges form in pairs, the practical form of representing hedges may be written as

$$\mu_U(x) = \mu(x)^{1/\alpha}$$
$$\mu_L(x) = \mu(x)^\alpha$$

with $\alpha \in [1, \infty)$.

1.11 Summary

This chapter discusses on IFSs, intuitionistic fuzzy operations and IFRs. It also discusses on interval-valued fuzzy set and Type II fuzzy set. IFS deals with the vagueness and also the uncertainty in describing the vagueness. The vagueness is represented in the form of a membership function, and the uncertainty is described in the form of hesitation degree while defining the membership function.

Type II fuzzy set considers the fuzzy membership function as fuzzy. This uncertainty is represented as lower and upper levels of the membership function. Sometimes, the membership degrees are not exactly defined but a vague range is defined and so the membership function lies in an interval range. This is the interval-valued fuzzy set.

Similar to the operations and relations on fuzzy sets, there are also operations and relations on IFSs that use *t*-norms and *t*-conorms, namely, are union, intersection, complement and negate. IFRs such as reflexivity, symmetricity, transitivity and equivalence relation are also discussed along with examples.

References

1. Atanassov, K.T., Intuitionistic fuzzy set, *Fuzzy Sets and Systems*, 20, 87–97, 1986.
2. Burillo, P. and Bustince, H., Intuitionistic fuzzy relations, *Mathware and Soft Computing*, 3(Part I), 5–38, 1995.
3. Bustince, H., Construction of intuitionistic fuzzy relations with predetermined properties, *Fuzzy Sets and Systems*, 109(3), 379–403, 2000.
4. Bustince, H. and Burillo, P., Structures of intuitionistic fuzzy relations, *Fuzzy Sets and Systems*, 78, 293–303, 1996.
5. Bustince, H., Kacpryzk, J., and Mohedano, V., Intuitionistic fuzzy generators: Application to intuitionistic fuzzy complementation, *Fuzzy Sets and Systems*, 114, 485–504, 2000.
6. Bustince, H. and Burillo, P., Correlation of interval-valued intuitionistic fuzzy sets, *Fuzzy Sets and Systems*, 74, 237–244, 1995.
7. Chaira, T., Enhancement of medical images using Atanassov's intuitionistic fuzzy domain using an alternative intuitionistic fuzzy generator with application to medical image segmentation, *International Journal of Intelligent and Fuzzy Systems*, 27(3), 1347–1359, 2014.
8. De, S.K., Biswas, R., and Roy, A.R., Some operations on intuitionistic fuzzy sets, *Fuzzy Sets and Systems*, 114, 477–484, 2000.
9. Ensafi, P. and Tizoosh, H.R., Type II fuzzy image enhancement, in *Lecture Notes in Computer Sciences*, Vol. 3656, M. Kamel and A. Campilho (Eds.), Springer, Berlin, Germany, pp. 159–166, 2005.

10. Klir, G.J. and Yuan, B., *Fuzzy Sets and Fuzzy Logic*, Prentice-Hall, Upper Saddle River, NJ, 1995.
11. Lei, Y.J., Wang, B.S., and Miao, Q.G., On the intuitionistic fuzzy relations with compositional operations, *Systems Engineering – Theory and Practice*, 25(2), 113–118, 2005.
12. Mendel, J.M. and John, R.I.B., Type 2 fuzzy sets made simple, *IEEE Transaction on Fuzzy Systems*, 10(2), 117–127, 2002.
13. Park, D.G. et al., Correlation coefficient of interval-valued intuitionistic fuzzy sets and its application to multiple attribute group decision making problems, *Mathematics and Computers Modelling*, 50, 1270–1293, 2009.
14. Roychowdhury, S. and Pedrycz, W., An alternative characterization of fuzzy complement functional, *Soft Computing*, Springer-Verlag, 7, 563–565, 2003.
15. Sugeno, M., Fuzzy measures and fuzzy integral: A survey, in *Fuzzy Automata and Decision Processes*, M.M. Gupta, G.S. Sergiadis, and B.R. Gaines (Eds.), North Holland, Amsterdam, the Netherlands, 89–102, 1977.
16. Yager, R.R., Some aspects of intuitionistic fuzzy sets, *Fuzzy Optimization and Decision Making*, 8(1), 67–90, 2009.
17. Yager, R.R., On the measures of fuzziness and negation. Part II: Lattices, *Information and Control*, 44, 236–260, 1980.
18. Zeng, W. and Li, H., Note on: Some operations on intuitionistic fuzzy sets, *Fuzzy Sets and Systems*, 157, 990–991, 2006.

2

Medical Image Processing

2.1 Introduction

The environment of medical imaging has changed dramatically by changing from analogue to digital technology. The widely used techniques are computed tomography (CT), Doppler imaging, magnetic resonance imaging (MRI), functional MRI and many others. The main task of image processing is converting the pictures into digital form. A diversity of problems in image processing range from image perception to image interpretation. Low-level image processing is concerned with the pixels, which initially enhances or filters the image, that is, it preprocesses the image. The next level of the image processing is concerned with the regions in the image. Here, the features of different regions or objects are extracted. The next process is high-level processing where the whole or parts of the image or series of the images are considered. The images are segmented for classification and recognition and finally interpretation. In the high level, the knowledge of experts is involved for making decisions. The following are the steps that are required in medical image processing.

2.1.1 Image Contrast Enhancement

The aim of image enhancement is to transform the original image into another image that is more suitable for further processing. It increases the contrast of the image by making the dark pixels darker and bright pixels brighter. As the images, especially the medical images, are not illuminated properly, many structures are not clearly visible and so contrast enhancement is required, which highlights the structures in the image. So, initially the images are enhanced and then they are processed, often called the preprocessing stage. In this stage, the images are filtered to remove any noise present, which is explained in detail in a different chapter.

2.1.2 Image Segmentation

Segmentation is one of the most important steps in medical image processing that is done normally after enhancement. It extracts any clot/abnormal lesion or blood cells/blood vessels present in an image. Each region in a segmented image possesses homogeneous properties with respect to features, such as grey level, texture or colour, and the property is different for different regions in an image. It partitions the image into disjoint sets corresponding to objects in the scene, which are very important in identifying different types of leukocytes or counting blood vessels or finding the size of the tumour or any other abnormalities present in the human body. After segmentation, analysis is carried out and is explained in detail in a separate chapter.

2.1.3 Boundary Detection

Boundary detection is an important process in medical image processing. It finds the structural information of the image, thus drastically reducing the data to be processed. In an image, there are changes in intensity levels. Edge is present when there is a change in grey level intensity. Edge image may be termed as gradient image. The boundaries indicate the location and the shape of the objects such as abnormal lesions or tumour or blood cells/vessels or any other structures present in an image. The edges in medical images are not distinct due to poor contrast and unequal illumination. So initially, the edges are sometimes enhanced before performing edge detection, which is explained in detail in a separate chapter.

2.1.4 Morphology

Morphology is a non-linear image processing technique that deals with the shape of the image features. Sometimes, there is a need to obtain a skeleton of an image or to dilate/erode an image or close/open an image to remove or fill any holes or gaps present in the image. It may be used to find the gradient of an image. It uses a structuring element of any size and shape which is applied on an image to perform operation. It is explained in detail in a separate chapter.

2.2 Image Registration

Image registration is the process of finding the transformation that aligns or maps an object of one coordinate to an object of a different coordinate. The goal of registration is to find a corresponding anatomical or functional position in two or more images. Maurer and Fitzpatrick [13] defined image registration as 'It is the one-to-one mapping between the coordinates in one space and those in another, such that the points in two spaces that correspond to the same anatomical point are mapped to each other'. Image registration is used mainly to detect the changes in same types of images. It is mainly used in remote-sensing applications such as in multi-spectral classification, environmental monitoring, change detection, weather forecasting and geographic information systems (GIS). In medical science, it is used in combining computer tomography and nuclear magnetic resonance (NMR) data to obtain complete information about the patient, monitoring tumour growth, comparing patient's data with anatomical atlases and many others.

Suppose ultrasound is taken at different times, and in order to observe any structural changes in two images, the mapping procedure is used. Detection is also required at different stages in tumour/lesion growth, before and after brain stimulus in functional MRI, rest and stress comparison and so on. Proper integration of data obtained from two images acquired in the clinical track of events is required, and this procedure is called registration. Sometimes, patients undergo many MR and CT at different times to study any changes in these images or x-ray time series to monitor the growth of specific bones or single-photon emission computed tomography (SPECT) to compare ictal and inter-ictal images and so on. After registration, fusion step is required to display the data. There are many image registration methods that may be classified in many ways. Maintz and Viergever [12] suggested a classification method based on nine criteria, which are classified into the following categories.

Dimensionality: It refers to the number of image dimensions involved, which may be 2D or 3D. The dimension can be spatial or time series, that is, time is added to it. Spatial dimension can be 2D–2D, 3D–3D or 2D–3D. Time series of images is that where time is added to the dimension. It is used in monitoring tumour growth or bone growth where images are taken at long intervals or post-operative monitoring of healing where images are taken at shorter intervals or evaluation of drug effects taken at various time intervals.

Nature of registration methods: Registration can be extrinsic where an artificial object is attached to the patient's body. The marker object used may be invasive or non-invasive, but non-invasive markers are less accurate. The following are examples of some markers: stereotactic frame that is screwed rigidly to the patient's outer skull used in neurosurgery [6], screw-mounted marker [3] and marker glued to the skin [11]. Registration can be intrinsic where the image information is generated by the patient. It can be as follows:

1. Landmark based where a limited set of identified points are identified and the accuracy depends on the accurate indication of corresponding landmarks in all modalities using some similarity measures.
2. Segmentation based where segmented structures (commonly object surfaces) are aligned. Segmentation based can be rigid where the same structures are extracted from both the images, and these are used as the input to the alignment process. It can be deformable

based where an extracted structure of one image is elastically deformed (deformable curves are active contours, snakes) in order to fit in the second image. Since segmentation is an important step, registration accuracy depends on the segmentation accuracy.

3. Voxel based where no segmentation is required and the methods directly depend on the grey values. It optimizes the similarity measure of all the geometrically corresponding voxel pairs for some features which can be only grey values.

Nature and domain of registration: The nature of transformation can be (1) rigid where the geometric transformation preserves all the distances (it includes two specifications – rotations and translational); (2) affine if it maps parallel lines onto parallel lines, that is, it preserves the straightness of lines and (3) projective if it maps lines onto lines and they may not preserve parallelism. Transformation is called elastic or curved if it maps lines onto curves. Transformation may be global if it is applied to the entire image and is called local if it is applied to the subsections of the image, with each subsection having its own transformation.

Interaction: It refers the control by human experts over the registration algorithm. Interaction can be either the initialization of certain parameters of the algorithm or, depending on the visual assessment of the alignment, adjustments that are done throughout the process. But no interaction is required if the algorithm is fully automated.

Optimization procedure: It is an approach in algorithmic registration where the quality of the registration is estimated continually during the registration process in terms of some functions and mapping between the images and the function in the optimization procedure is either minimized or maximized.

Modalities involved in registration: It refers to the means by which the images to be registered are acquired. It can either be (1) monomodal, (2) multi-modal or (3) modality to modal and patient to modality. In monomodal registration, the images to be registered belong to the same modality, for example, in growth monitoring of tumour/cyst and ictal or inter-ictal comparison. In multi-modal registration, the images to be registered arise from two different modalities such as positron emission tomography (PET) to magnetic resonance (MR) image or MR to PET image. In modality to modal and patient to modality registration, one image is involved and the other 'modality' is either patient or modal.

Subject: Registration can be done (1) when all the images involved are acquired from a single patient, that is, from like modalities, it is called intra-subject registration, or (2) when two images are acquired from different patients or patient, often called inter-subject registration. Registration can also be carried out when one image is acquired from a single patient and the other image is constructed from an image information database obtained from the images of many subjects. It is called atlas registration. Normally, the registration of a patient image to an image of a normal patient image is referred to as atlas registration.

Methods based on neural networks, genetic algorithm and fuzzy sets are being used in medical image registration. In different computational aspects of registration, a neural network such as Hopfield network and self-organizing map is used. Genetic algorithm is used to find the exact or approximate solutions to optimization and search procedure.

Majority of image registration methods involve the following steps:

Feature detection: In this step, features of distinctive objects such as boundary, edges, contour and corner are manually or automatically detected.

Feature matching: In this step, matching between the features detected in the sensed image and reference image is computed. Various feature descriptors and similarity measures along with spatial relationships among the features are used for such purpose. In feature matching, due to different imaging conditions, features cannot be similar. So, the choice of feature descriptors and similarity measures is considered critically.

Model transformation: In this step, parameters that are used in the mapping function or in aligning the sensed image with the reference image are estimated. The parameters are computed by means of feature correspondence. While mapping, prior information about the acquisition process is also required.

Image transformation: The mapping function that is constructed is used to transform the sensed image to register the images.

Fuzzy sets are widely used that considers the uncertainty in the data. It is a mathematical tool which is used in the registration process to find registration transformation or to preprocess the image to remove noise and enhance the features using image gradient or gamma correction before segmenting/clustering the regions of the images [7,9,10] that are to be registered. The basic principle in most of the image registration algorithms is to find the alignment between the two images by maximizing the fuzzy similarity. Again, in a registration problem as, for example, in a feature-based registration problem, fuzzy c means clustering (FCM) is used to detect the regions defined by the users to find a match between the segments/regions in reference and target images. FCM is used in feature space where features of each pixel are considered. In fuzzy image registration, different fuzzy methods are used for obtaining better results and these methods can be found in many books. Again these fuzzy methods can be extended using intuitionistic fuzzy set and Type II fuzzy set theories for better registration than fuzzy methods where more uncertainties are considered.

2.3 Image Fusion

Image fusion is a process of combining information of two or three images to form a single image or combine some of their features in a single image. It is used to improve the imaging quality and to reduce the randomness and redundancy so that the assessment of medical problems becomes easier. Its aim is to combine redundant information from multiple images to create a fused image. The new image so generated contains a more accurate description of the image than a source image and is more suitable for human visual or further image processing and analysis tasks. For medical images, fusion can lead to additional clinical information that may not be apparent in the separate images and can also reduce storage cost by storing the single fused image instead of multi-source images. Image fusion methods can be grouped into three categories: pixel level, feature level and decision level. In the pixel level, the simplest way is to take the average of two images pixel by pixel, but many undesirable effects are observed. Many other techniques are available such as weighted average, principal component analysis (PCA) and Brovey transform [18,22]. In the feature level, the features involved are edges, regions, shape, size, length or image segments. These features are then combined with the similar features present in other input images to form the final fused image.

2.4 Image Retrieval

In medical image retrieval [5,15], the number of digitally produced images is rising enormously. Thus, more efficient image retrieval methods are required for better management of medical information system. There are two methods that the medical images are retrieved; these are text-based and content-based methods [8] or a combination of the two. In the text-based retrieval system, images are retrieved by manually annotated text description and traditional database techniques to manage the images. It works fast and is reliable when the images are well annotated, but it does not work better on un-annotated image database. Also, the annotation procedure is time consuming and commonly results in irrelevant images. In content-based retrieval, images are retrieved and indexed based on their visual features such as colour, texture and shape. Commonly used features are colour features such as red, green and blue (RGB); hue, saturation and value (HSV); CIELab; and Luv. But as most of the medical images are in greyscale, colour features are not used in medical image retrieval. Textural features mean the spatial organization of pixels in the images and find the characteristics of the image in a certain direction. Features can be obtained using Fourier transform, Gabor filters, wavelets, co-occurrence matrices and many others. In medical images, textural features are required as they reflect the details of the image structure. Shape features include edges, boundaries, contour, curve and surfaces.

Content-based image retrieval is a system for browsing, searching and retrieving images from a large database. It is a problem of searching for digital images from the large database where images are retrieved based on image features. In this retrieval system, a query image is required and is allowed to match the images in the database based on some similarity measures. Image features that may be colour or texture or shape for both the query image and the images in the database are used in matching. In the matching procedure, either the entire image is used or subimages are used depending on the user's choice. For each image in the database, image features are computed and likewise features of the query image are also computed. Using the similarity measure, a match is found between the query image and the images in the database.

Management and access of these medical images become very complex. Access is mainly based on patient identification or study characteristics. The purpose of a medical image retrieval system is to deliver the needed information at the right time and to right person to improve the quality of the care process. In the decision-making process, the retrieval system will find other images of the same modality and same anatomical region of the same disease.

Storing and accessing these large numbers of images are very important. As it requires large space, data reduction techniques are used. For feature reduction, PCA is used, which is also called Karhunen–Loeve transform (KLT).

2.5 Fuzzy Processing of Medical Images

Medical image processing may be done using crisp or fuzzy set theory. Medical image processing is intended as a central resource for information of image processing in the medical field. Classical methods based on crisp set theory are mentioned in many texts. In crisp set, the structures present in the image are considered to have a rigid demarcating boundary and crisp methods consider bivalent logic, and the degree of belongingness (membership degree) of a pixel present in an image is either 0 or 1. But the objects in the real world do not always have a rigid demarcating boundary and so there is a gradual transition of the membership degree from zero to unity membership, and this is what fuzzy set theory deals with as introduced by Zadeh [24]. It considers one uncertainty, which is the degree of belongingness of an element in a set. The application of fuzzy set theoretic concepts in image processing took formal shape only in the 1980s with the pioneering research carried out by Pal et al. [16,17] and Rosenfeld and Pal [20,21].

With the emergence of fuzzy set and its significance, it is widely used in diverse applications such as medical image processing, remote-sensing application, pattern recognition and decision-making problems. The raw medical images are blurred and unsharp with vague structures and unclear boundaries. Bezdek [2] used fuzzy models for segmentation and edge detection in medical image data. Images are in a sense 'fuzzy', and the task is to reduce the fuzziness present in the image to obtain a clear image with sharp boundaries. So, the task of medical image processing is not only to fuzzify the image (i.e. to consider the vagueness present) but also to defuzzify the image (i.e. to obtain a clear image).

While attempting to process medical images for analysis, it may be a good idea to consider the fact that a computer vision system is usually embedded with uncertainty and vagueness which needs to be appropriately taken care of. Let us now see the logical reasons behind the applicability of fuzzy notions in image processing by Prewitt [19]:

1. Images are 2D projections of the objects in a 3D world, and so some information is lost while mapping.

2. In grey-level images, uncertainty exists due to the variability of grey values. The pixel values in digital images are considered imprecise and should be viewed as fuzzy numbers.

3. Many of the image definitions, such as boundaries/regions of objects or the homogeneity of the segments or the contrast between the objects and the background, require fuzzy notions for their characterization.

4. Ambiguity is often present in the final interpretation of the image. Since human understanding is never crisp or precise, it is important to incorporate soft decision-making since a hard decision may often be costly.

Fuzzy image processing is a collection of different fuzzy approaches to image processing that can understand, represent and process the images. It has three main stages – fuzzification (membership plane), modification of membership values and, finally, defuzzification – as shown in Figure 2.1.

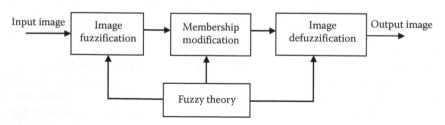

FIGURE 2.1
The general structure of fuzzy image processing.

The main part is fuzzification where the grey levels are transformed to the membership function. The membership function is the degree of belongingness of the pixels in an image. After fuzzification, depending on the user's requirements, the membership values of the grey level are modified using an appropriate fuzzy technique.

2.6 Advanced Fuzzy Processing of Medical Images

Till now we know that uncertainties in an image can be dealt better with fuzzy set theory. But with the introduction of advanced fuzzy set theories where more or different types of uncertainties are considered, researches are carried out for obtaining accurate results on medical images. In this section, advanced fuzzy techniques such as intuitionistic fuzzy sets and Type II fuzzy sets are explored for medical image processing.

2.6.1 Intuitionistic Fuzzy Set

As medical images are not equally and well illuminated, for a more accurate diagnosis, advanced fuzzy set theoretic techniques that include intuitionistic fuzzy set and Type II fuzzy set are being explored in recent days. Fuzzy set theory performs better as it considers the vagueness in the form of a membership function. The membership function is not properly defined for a particular image. It may be Gaussian, gamma, triangular or any other function. This is the reason different researchers get varied results with different membership functions. Membership function selection depends on the user's choice. So, some kind of hesitation or uncertainty is present while defining the membership function. This hesitation is taken into account in the intuitionistic fuzzy set theory introduced by Atanassov [1]. In fuzzy set, the membership degrees lie between 0 and 1 and the non-membership degree is the complement of the membership degree. But in intuitionistic fuzzy set, due to the hesitation degree, the non-membership degree is not the complement of the membership degree; rather, it is less than or equal to the complement of the membership degree. It offers an improved performance as compared to fuzzy set theory. So, in intuitionistic fuzzy set, two uncertainties are considered: the membership degree and non-membership degree. With the consideration of more uncertainties, attempts have been taken by several researchers to work on medical images. As medical images contain uncertainties, intuitionistic fuzzy set may be useful in medical image processing. The reason is simple: In fuzzy set, only the membership degrees are considered; non-membership values are not considered as they are the complement of the membership degree. An intuitionistic fuzzy set A in X may be represented as

$$A = \{(x, \mu_A(x), \nu_A(x), \pi_A(x)) \mid x \in X\}$$

with the condition

$$\pi_A(x) + \mu_A(x) + \nu_A(x) = 1$$

It is obvious that

$$0 \le \pi_A(x) \le 1, \quad \text{for each } x \in X$$

2.6.2 Type II Fuzzy Set

Again, if we explore properly, we find that the membership function defined in the ordinary or type I fuzzy set theory is imprecise and uncertain. The membership function in type I fuzzy set is considered to be 'fuzzy', and thus the new fuzzy set is named as Type II fuzzy set, introduced by Zadeh [23]. Type II fuzzy set accounts this uncertainty by considering another degree of freedom for better representation of uncertainty where the membership functions are themselves fuzzy [14]. So, if the membership function of type I fuzzy set is blurred, then Type II fuzzy set is obtained. Type I fuzzy set can be defined by assigning upper and lower membership degrees to each element, and the membership function does not have a single value for every element but an interval – lower and upper intervals – which are written as

$$\mu^{\text{upper}} = [\mu(x)]^{\alpha}$$
$$\mu^{\text{lower}} = [\mu(x)]^{1/\alpha} \tag{2.1}$$

where $\alpha \in [0, 1]$. So, a more practical form of representing Type II fuzzy set is written as

$$A_{\text{TypeII}} = \{x, \mu_U(x), \mu_L(x) \mid x \in X\}$$

and

$$\mu_L(x) < \mu(x) < \mu_U(x), \quad \mu \in [0,1]$$

So, in Type II fuzzy set, the membership function lies in an interval, and this may be very useful in medical image processing where the fuzziness is present in an interval.

The main reason to carry out research using advanced fuzzy set theory is to obtain better and accurate results for better diagnosis.

2.7 Some Applications of Advanced Fuzzy Set in Medical Image Processing

Fuzzy image analysis finds applications in many computer vision systems including robot vision, object recognition, biomedical image processing and remotely sensed scene analysis. With the introduction of intuitionistic fuzzy set and Type II fuzzy set, researches are carried out on medical images for obtaining better results. However, applying a single image analysis method may not always yield reliable results. It is observed that when an ensemble of image processing techniques is applied to an image, appropriate results are likely to be obtained. The raw medical image that is initially obtained does not have better contrast. So, the image is required to enhance before going to another process. Next, segmentation is required before analysis. Segmentation divides the abnormal lesions or haemorrhage/clots or blood vessels and different types of leukocytes. After segmentation, analysis is done.

A few applications of advanced fuzzy set theory in medical image processing are discussed:

1. *Image contrast enhancement*: There are many image enhancement methods, and the most common is histogram equalization. But this method may not always yield good results on medical images, since pixel grey levels in an image are imprecise. To deal with such kind of images, many authors suggested fuzzy methods to handle the inexactness of grey values. The most common operator is the contrast intensification (INT) operator, which is applied globally to modify the membership and increase the contrast of the image. This approach transforms the higher membership values to much higher values or lower membership values to much lower values in a non-linear manner. But this method has some limitation, which is improved using the NINT operator that uses Gaussian-type fuzzification function. In many cases, global histogram approaches to enhancement fail to achieve satisfactory results as these methods consider the whole image in its totality, so local enhancement techniques work better. Enhancement using fuzzy methods performs well as it considers the vagueness in the image, but in some cases, proper enhancement is not achieved using the fuzzy method. Then intuitionistic and Type II fuzzy set theories are used that use more uncertainties or different types of uncertainty. This is done to

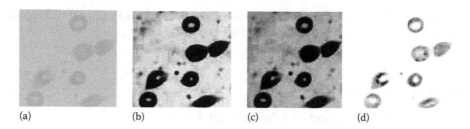

FIGURE 2.2
(a) Blood vessel image, (b) fuzzy NINT operator method, (c) intuitionistic fuzzy method and (d) Type II fuzzy method.

obtain better results. So, when better results are not obtained using fuzzy, then advanced fuzzy techniques may be used. An example in Figure 2.2 shows contrast enhancement using fuzzy, intuitionistic fuzzy and Type II fuzzy methods.

2. *Image segmentation*: In many cases, image region boundaries may not have a sharp transition and so fuzzy decision as to whether the pixel belongs to a region is used. But in many cases, fuzzy set theory does not segment the images clearly. In that case, intuitionistic fuzzy set or Type II fuzzy set may be used for obtaining better thresholded images. As intuitionistic fuzzy set considers more uncertainties and Type II fuzzy set considers different types of uncertainty as compared to fuzzy set, better results may be obtained. An example in Figure 2.3 shows thresholded images using fuzzy, intuitionistic and Type II fuzzy methods. The image shows abnormal leukocytes in the blood.

3. *Clustering*: It gathers similar pixels in a group and different pixels in different groups. This is useful in images where different regions of different intensities are present. In hard clustering, each pixel either belongs to a group or belongs to different groups, so the membership values of each pixel with respect to that group is 1 and with respect to different groups is 0. In fuzzy clustering, the pixel that belongs

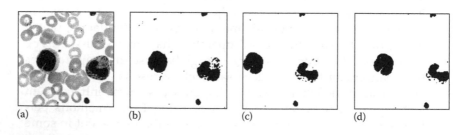

FIGURE 2.3
(a) Abnormal leukocyte cell images, (b) threshold using fuzzy set, (c) threshold using intuitionistic fuzzy set and (d) threshold using interval Type II fuzzy set.

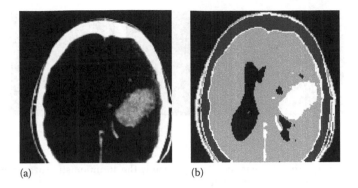

(a)   (b)

FIGURE 2.4
(a) Brain image 'tumour' and (b) intuitionistic fuzzy cluster.

to a group may contribute to different groups. So, the membership values with respect to different groups are not 0. Intuitionistic fuzzy clustering is also similar to fuzzy clustering; the only difference is that the membership function is updated with the hesitation degree. An example in Figure 2.4 shows fuzzy, intuitionistic fuzzy and Type II fuzzy clustering.

4. *Edge detection*: Edges/boundaries of abnormal lesions/tumour or any other abnormalities are required in image analysis. As medical images are poorly illuminated, the crisp decision as to whether an edge is present in the image is very difficult. So initially edges are sometimes enhanced before performing edge detection. The fuzzy method may be very useful in alleviating such type of problem. There are many fuzzy edge detection techniques, but these techniques may not find better edges in all the images. In that case, intuitionistic fuzzy set and Type II fuzzy set may be useful where more or different types of uncertainties are considered. An example in Figure 2.5 shows fuzzy edges using interval-valued fuzzy set and Type II fuzzy set.

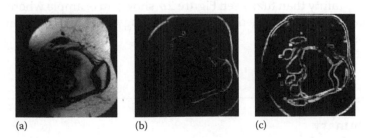

(a)   (b)   (c)

FIGURE 2.5
(a) Knee patella image, (b) edge image using the interval-valued fuzzy set and (c) edge image using the Type II fuzzy method.

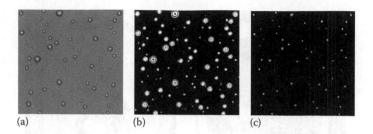

FIGURE 2.6
(a) Live cell image, (b) edge-enhanced image using the intuitionistic fuzzy method and (c) enhanced image using the fuzzy method.

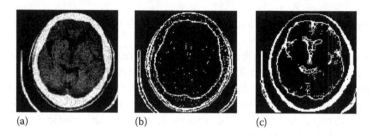

FIGURE 2.7
(a) Noisy CT scan brain image, (b) edge image using the fuzzy method and (c) edge image using the intuitionistic fuzzy approach.

An example in Figure 2.6 shows the edge-enhanced image using fuzzy and intuitionistic fuzzy sets.

5. *Morphology*: Morphology is a non-linear image processing technique that deals with the shape of the image features. Fuzzy mathematical morphology is almost a newly developed morphology by Bloch and Maitre [4] where fuzzy techniques are used. Later on, intuitionistic fuzzy concepts are also used to obtain better results. When in some cases fuzzy methods do not perform well, intuitionistic fuzzy methods may be useful as intuitionistic fuzzy set considers more uncertainty than fuzzy set. Figure 2.7 shows an example where edge detection is performed using morphology utilizing dilation and erosion.

2.8 Summary

In this chapter, an attempt has been made to discuss medical image processing and the use of fuzzy and advanced fuzzy set theories (intuitionistic fuzzy and Type II fuzzy sets) in medical image processing. The reason for using advanced fuzzy set theories is also discussed. Image registration, fusion and retrieval are also discussed in detail. Several applications using intuitionistic fuzzy and Type II fuzzy sets on enhancement, segmentation, clustering and morphology have been highlighted in this chapter with some examples and are discussed in detail in separate chapters of this book.

References

1. Atanassov, K.T., *Intuitionistic Fuzzy Sets, Theory and Applications*. Series in Fuzziness and Soft Computing, Physica-Verlag, Heidelberg, 1999.
2. Bezdek, J.C. et al., Medical image analysis with fuzzy models, *Statistical Methods in Medical Research*, 6, 191–214, 1997.
3. Ellis, R.E., Toksvig-Larsen, S., Marcacci, M., Caramella, D., and Fadda, M., A biocompatible fiducial marker for evaluating the accuracy of CT image registration, in *Computer Assisted Radiology*, Vol. 1124, Excerpta Medica – International Congress Series, Amsterdam, the Netherlands, pp. 693–698, 1996.
4. Bloch, I. and Maitre, H., Fuzzy mathematical morphologies: A comparative study, *Pattern Recognition*, 28(9), 1341–1387, 1995.
5. Hwang, K.H., Lee, H., and Choi, D., Medical image retrieval: Past and present, *Healthcare Informatics Research*, 18(1), 3–9, 2002.
6. Lemieux, L. and Jagoe, R., Effect of fiducial marker localization on stereotactic target coordinate calculation in CT slices and radiographs, *Physics in Medicine and Biology*, 39, 1915–1928, 1994.
7. Lu, Z. et al., Image registration based on fuzzy similarity, in *IEEE/ICME International Conference on Complex Medical Engineering*, Beijing, China, pp. 517–520, 2007.
8. Liu, Y. et al., A survey of content based retrieval with high level semantics, *Pattern Recognition*, 40, 262–282, 2007.
9. Maes, F. et al., Multimodality image registration by maximization of mutual information, *IEEE Transactions on Medical Imaging*, 16(2), 187–198, 1997.
10. Mahmoud, H., Masulli, F., and Rovetta, S., Feature-based medical image registration using a fuzzy clustering segmentation approach, in *Computational Intelligence Methods in Bioinformatics and Biostatistics*, Vol. 7845, L.E. Peterson, F. Masulli, and G. Russo (Eds.), Springer-Verlag, Berlin, Germany, pp. 37–47, 2013.
11. Malison, R.T. et al., Computer-assisted coregistration of multislice SPECT and MR brain images by fixed external fiducials, *Journal of Computer Assisted Tomography*, 17(6), 952–960, 1993.
12. Maintz, J.B.A. and Viergever, M.A., A survey of medical image registration, *Medical Image Analysis*, 2(1), 1–36, 1998.
13. Maurer, C.R. and Fitzpatrick, J.M., A review of medical image registration, in *Interactive Image-Guided Neurosurgery*, R.J. Maciunas (Ed.), American Association of Neurological Surgeons, Park Ridge, IL, pp. 17–44, 1993.
14. Mendel, J.M. and John, R.I.B., Type 2 fuzzy sets made simple, *IEEE Transactions on Fuzzy Systems*, 10(2), 117–127, 2002.

15. Muller, H. et al., A review of content-based image retrieval systems in medical applications – Clinical benefits and future directions, *International Journal of Medical Informatics*, 73, 1–23, 2004.
16. Pal, N.R., Pal, S.K., and Bezdek, J.C., A mixed C-means clustering model, in *Proceedings of the FUZZ-IEEE-97*, Barcelona, Spain, pp. 11–21, 1997.
17. Pal, S.K., King, R.A., and Hishim, A., Automatic gray level thresholding through index of fuzziness and entropy, *Pattern Recognition Letters*, 1, 141–146, 1983.
18. Pradhan, P.S., King, R.L., and Younan, N.H., and Holcomb, D.W., Estimation of the number of decomposition levels for a wavelet-based multiresolution multisensor image fusion, *IEEE Transactions on Geoscience and Remote Sensing*, 44(12), 3674–3686, 2006.
19. Prewitt, J.C., Object enhancement and extraction, in *Picture Processing and Psychopictorics*, B.S. Lipkin and A. Rosenfeld (Eds.), Academic Press, New York, pp. 75–149, 1970.
20. Rosenfeld, A., The fuzzy geometry of image subsets, *Pattern Recognition Letters*, 2, 311–317, 1984.
21. Rosenfeld, A. and Pal, S.K., Image enhancement and thresholding by optimization of fuzzy compactness, *Pattern Recognition Letters*, 7, 77–86, 1984.
22. Yang, Y. et al., Medical image fusion via an effective wavelet-based approach, *EURASIP Journal on Advances in Signal Processing*, 2010, 1–13, 2010.
23. Zadeh, L.A., The concept of a linguistic variable and its application to approximate reasoning-I, *Information Sciences*, 8, 199–249, 1975.
24. Zadeh, L.A., Fuzzy sets, *Information and Control*, 8, 338–353, 1965.

3

Fuzzy and Intuitionistic Fuzzy Operators with Application in Decision-Making

3.1 Introduction

With the demand for knowledge handling systems capable of dealing with imprecision, a formal mathematical tool is quiet necessary. In real decision-making, decision-making problems are fuzzy and uncertain, and the attribute values are not expressed always as real numbers; rather, some of them are more suitable to be denoted as fuzzy numbers. The theory of fuzzy logic provides a mathematical strength to capture the uncertainties and is suitable to process human information and reasoning. Fuzzy logic operators are used to decide multi-criteria decision-making problem. Decision-making is defined as making choices between future and uncertain alternatives. It is a difficult process due to incomplete and imprecise information, vagueness and uncertainty of the situation. These factors show that decisions can take place in a fuzzy logic environment.

Aggregating criterion functions to form decision functions is of great importance in all disciplines. At one extreme of the situation, we desire that all the criteria are fulfilled, and in another extreme, we desire that any of the criteria is fulfilled. These two extremes lead to the use of 'and' and 'or' operators to combine the criterion function.

3.2 Fuzzy Operators

Since the introduction of fuzzy sets by Zadeh, fuzzy set theory is extended from ordinary set theory and the operators in ordinary set theory can be extended to operators that are capable of realization of the same union, intersection and complement logical operators in a multi-valued logic. In propositional logic, the operators OR, AND and NOT are used to build new propositions.

If A and B are two fuzzy sets and $\mu_A(x)$ and $\mu_B(x)$ are the membership values of the elements, the union operator can be constructed using the OR operator and is defined as

Union operator: $\mu_{A \cup B}(x) = \mu_A(x) \vee \mu_B(x)$

Intersection operator: $\mu_{A \cap B}(x) = \mu_A(x) \wedge \mu_B(x)$

Complement: $\mu_{A^c}(x) = 1 - \mu_A(x)$

These operators fulfil the following axioms:
OR operator

1. Monotonicity: If $a < a'$ and $b < b'$, then $a \vee b < a' \vee b'$
2. Commutativity: $a \vee b = b \vee a$
3. Boundary condition: $0 \vee 0 = 0, 0 \vee 1 = 1, 1 \vee 0 = 1, 1 \vee 1 = 1$
4. Associativity: $(a \vee b) \vee c = a \vee (b \vee c)$

The AND operator follows similar types of axioms with the following boundary condition:
$0 \vee 0 = 0, 0 \vee 1 = 0, 1 \vee 0 = 0, 1 \vee 1 = 1$

Complement operator: $\bar{a}$

Boundary condition: $\bar{1} = 0, \bar{0} = 1$

Monotonicity: If $a \leq b$, then $\bar{b} \leq \bar{a}$

Involutive: $\overline{(\bar{a})} = a$

Fuzzy logic operators are widely used in fuzzy inference systems where if–then rules follow. Suppose there are two fuzzy sets A with $A = \{a_1, a_2, ..., a_n\}$ and B with $B = \{b_1, b_2, ..., b_n\}$, the rules may be written as:

R_1: If $(a \wedge b)$, then c

R_2: If $(a \vee b)$, then c

These fuzzy operators can be written using the min($\wedge$) and max($\vee$) functions.

3.3 Fuzzy Operators Induced by Fuzzy *t*-Norm and *t*-Conorm

Suppose there are $A_1, A_2, A_3, ..., A_n$ criteria in multi-criteria decision-making problem. For each A_j, let $A_j(x) \in [0, 1]$ be a degree to which x satisfies the criteria. The problem is to formulate an overall decision D such that for any x, $D(x) \in [0, 1]$ denotes the degree to which x satisfies the desired requirement. The problem becomes

$$D(x) = F\{A_1(x), A_2(x), A_3(x), \ldots, A_n(x)\}$$

Suppose we desire to model the interrelationship between the criteria. At one extreme, one can desire that 'all' the criteria are satisfied. So, x must satisfy A_1 and A_2 and A_n. Thus, the requirement is 'anding' the values. At another extreme, we may desire that at least one of the criteria is satisfied. So, x must satisfy A_1 or A_2 or A_n. Thus, the requirement is 'oring' the values.

There exist a class of operators called t-norms that quantitatively implement 'anding' aggregation which implies 'all' the criteria are satisfied. Similar to t-norms, there also exist t-conorms where 'oring' aggregation is applied which implies that 'at least' one of the criteria is satisfied. Recently, many authors proposed t-norms and t-conorms. t-norm serves as a union operator, and t-conorm serves as an intersection operator which can be used to handle multiple rules. In this section, different t operators are discussed along with their properties.

3.3.1 *t*-Norm

A t-norm T: $[0, 1] \rightarrow [0, 1]$ is a kind of binary operation used in the framework of fuzzy logic and probabilistic metric space. It represents the intersection in fuzzy set theory or an 'ANDing' operator. The four basic t-norms are

1. Zadeh's intersection: $T_M(x, y) = \min(x, y)$
2. Product intersection: $T_P(x, y) = x \cdot y$
3. Lukasiewicz intersection: $T_L(x, y) = \max(x + y - 1, 0)$
4. Nilpotent T-norm: $T_n(x, y) = \min(x, y)$,   if $x + y > 1$
   $\qquad\qquad\qquad\qquad\quad 0$,   otherwise

Definition: T: $[0, 1] \times [0, 1] \rightarrow [0, 1]$ is a t-norm iff it satisfies the following properties where $x, y, z \in [0, 1]$ [12,15]:

1. Boundary condition: $T(0, 0) = T(0, 1) = T(1, 0) = 0$ and $T(1, 1) = 1$.
2. Associativity: $T(T(x, y), z) = T(x, T(y, z))$.
3. Commutativity: $T(x, y) = T(y, x)$.
4. Monotonicity: $T(x, y) \le T(x, z)$ if $y \le z$.
5. One identity: $T(x, 1) = x$.
6. A t-norm, T, is called Archimedean iff.
7. T is continuous.
8. $T(x, x) < x$ for all $x \in (0, 1)$.

A function T: $[0, 1] \times [0, 1] \to [0, 1]$ is called an Archimedean t-norm iff there exists a decreasing and continuous function f: $[0, 1] \to [0, \infty]$ with $f(1) = 0$:

$$T(x,y) = f^{-1}(f(x) + f(y)) \tag{3.1}$$

3.3.2 *t*-Conorm

A t-conorm T^*: $[0, 1] \to [0, 1]$ is a kind of binary operation used in the framework of fuzzy logic and probabilistic metric spaces. It represents union in fuzzy set theory or an 'ORing' operator. The four basic t-conorms are

1. Zadeh's union: $T_M^*(x,y) = \max(x,y)$
2. Product union: $T_P^*(x,y) = x + y - x \cdot y$
3. Lukasiewicz union: $T_L^*(x,y) = \min(x+y,1)$
4. Nilpotent T-conorm: $T_n^*(x,y) = \max(x,y), \quad$ if $x + y < 1$
   $\qquad\qquad\qquad\qquad\quad 0, \qquad\qquad$ otherwise

Definition: T^*: $[0, 1] \times [0, 1] \to [0, 1]$ is a t-conorm iff it satisfies the following properties where $x, y, z \in [0, 1]$ [12,15]:

1. Boundary condition: $T^*(0, 0) = 0$ and $T^*(0, 1) = T^*(1, 0) = T^*(1, 1) = 1$.
2. Commutativity: $T^*(x, y) = T^*(y, x)$.
3. Monotonicity: $T^*(x, y) \le T^*(x, z)$ if $y \le z$.
4. Associativity: $T^*(T^*(x, y), z) = T^*(x, T^*(y, z))$.
5. Zero identity: $T^*(x, 0) = x$.
6. A T-conorm, T^*, is called Archimedean iff.
7. T^* is continuous.
8. $T^*(x, x) > x$ for all $x \in (0, 1)$.

A function T^*: $[0, 1] \times [0, 1] \to [0, 1]$ is called an Archimedean t-conorm iff there exists an increasing and continuous function g: $[0, 1] \to [0, \infty]$ with $g(0) = 0$:

$$T^*(x,y) = g^{-1}(g(x) + g(y)) \tag{3.2}$$

f and g are additive generators.

There are many triangular operators, and they are classified as a class of (1) conditional operators and (2) algebraic operators [7,12,15,18]. Conditional operators contain min or max operators or the combination of min and max operators. *t*-norms and *t*-conorms that belong to the conditional class are as follows:

1. Yager's *t*-norm [18]: $Y_p(x,y) = 1 - \min([(1-x)^p + (1-y)^p]^{(1/p)}, 1)$

   with decreasing generator $f_p(x) = (1-x)^p$

   with $f^{-1}(y) = 1 - y^{1/p}$, $p > 0$

   T-conorm: $Y_p^*(x,y) = \min([x^p + y^p]^{(1/p)}, 1)$ (3.3)

   with increasing generator

   $g_p(x) = x^p$ and $g_p^{-1}(y) = y^{1/p}$

2. Zadeh's *t* operators:

$$T(x,y) = \min(x,y)$$
$$T^*(x,y) = \max(x,y)$$

    (3.4)

3. Weber's operator [15]:

   T-norm: $W(x,y) = \max\left(\dfrac{x+y-1+\lambda xy}{1+\lambda}, 0\right)$, $\lambda > -1$

   with decreasing generator $f(x) = 1 - \dfrac{\ln(1+\lambda x)}{\ln(1+\lambda)}$

   with $f^{-1}(y) = \dfrac{1}{\lambda}[(1+\lambda)^{1-y} - 1]$, $y \leq 1$

   T-conorm: $W^*(x,y) = \min(x+y+\lambda xy, 1)$ (3.5)

   where the increasing generator is $g(x) = \dfrac{\ln(1+\lambda x)}{\ln(1+\lambda)}$

   with $g^{-1}(y) = \dfrac{1}{\lambda}[(1+\lambda)^y - 1]$, $y \leq 1$

4. Schweizer and Sklar's [13] *t*-norm and *t*-conorm:

   $S(x,y) = [\max(0, x^r + y^r - 1)]^{1/r}$

   $S^*(x,y) = 1 - [\max(0, (1-x)^r + (1-y)^r - 1)]^{1/r}$, $r > 0$

    (3.6)

5. Dubois and Prade [5]:

$$T\text{-norm:} \quad D(x,y) = \frac{xy}{\max(x,y,c)}$$

$$T\text{-conorm:} \quad D^*(x,y) = \frac{x+y-xy-\min(x,y,1-c)}{1-\min(x,y,1-c)} \quad \text{and } c \in (0,1)$$

(3.7)

6. Sugeno's *t*-norm and *t*-conorm [14]:

$$T(x,y) = \max(0,(1+b)(x+y-1)-bxy)$$

$$T^*(x,y) = \min(1, x+y-bxy), \quad b > -1$$

(3.8)

Algebraic operators do not contain conditional operators and are purely algebraic in nature. This class of operators does not have conditional function; rather, they simply follow arithmetic operations:

1. Chaira suggested an algebraic *t*-norm and *t*-conorm that does not contain min or max operators. The following shows the calculation for finding *t*-norms and *t*-conorms:

$$g(x) = \ln\left(\frac{1+(\lambda+1)x}{1-x}\right), \quad 0 \le \lambda \le 1$$

where $g(\cdot)$ is an increasing function such that $g(0) = \ln(1) = 0$ with

$$g^{-1}(x) = \frac{e^x - 1}{e^x + 1 + \lambda}$$

As x approaches 1, the limit of the logarithm value tends to be infinity. That means it is an increasing function:

$$C^*(x,y) = g^{-1}(g(x)+g(y)), \quad g(x)+g(y)$$

$$= g^{-1}\left[\ln\left(\frac{1+(\lambda+1)x}{1-x}\right) + \ln\left(\frac{1+(\lambda+1)y}{1-y}\right)\right]$$

$$C^*(x,y) = \frac{1+(1+\lambda)x+(1+\lambda)y+(1+\lambda)^2xy-(1-x)(1-y)}{1+(1+\lambda)x+(1+\lambda)y+(1+\lambda)^2xy+(1-x)(1-y)+\lambda(1-x)(1-y)}$$

$$= \frac{2x+2y+\lambda x+\lambda y+2\lambda xy+\lambda^2xy}{2+2xy+3\lambda xy+\lambda^2xy+\lambda} = \frac{(\lambda+2)(x+y+\lambda xy)}{(\lambda+2)(1+xy+\lambda xy)}$$

$$= \frac{x+y+\lambda xy}{(1+\lambda)xy+1}, \qquad\qquad (3.9)$$

From this equation, it is seen that

$$C^*(0,0) = \frac{0}{1}$$

$$C^*(1,1) = \frac{1+1+\lambda}{1+\lambda+1} = \frac{2+\lambda}{2+\lambda} = 1$$

$$C^*(1,0) = \frac{1}{1} = 1$$

$$C^*(x,0) = \frac{x+0+0}{1} = x$$

The *t*-conorm is strict and satisfies the criteria.

T-norm: The generating function for *t*-norm is given by

$$f(x) = \ln\left(\frac{1+(1-\lambda)(1-x)}{x}\right), \quad 0 \le \lambda \le 1$$

which is a decreasing function where $f(1) = \ln((1 + 0)/1) = 0$. As x approaches 0, the limit of the logarithm tends to be infinity as $0 \le \lambda \le 1$. That means it is a decreasing function and $f^{-1}(x) = (2 - \lambda)/(e^x + 1 - \lambda)$:

$$C(x,y) = f^{-1}(f(x)+f(y))$$

$$= \frac{(2-\lambda)xy}{[1+(1-\lambda)(1-x)][1+(1-\lambda)(1-y)]+(1-\lambda)xy}$$

$$= \frac{xy}{(\lambda-1)(x+y-xy)+(2-\lambda)} \qquad\qquad (3.10)$$

From this equation, it is seen that

$$C(0,0) = 0$$

$$C(1,1) = \frac{1}{\lambda - 1 + 2 - \lambda} = 1$$

$$C(1,0) = C(0,1) = 0$$

Also $C(x, 1) = x$, implying that the *t*-norm is strict.

2. Hamacher [7] proposed *t*-norm and *t*-conorm that do not contain min or max operators:

$$H_\gamma(x,y) = \frac{x \cdot y}{\gamma + (1-\gamma) \cdot (x + y - x \cdot y)}, \quad \gamma > 0 \tag{3.11}$$

is a *t*-norm with decreasing generator as

$$f_\gamma(x) = \frac{1}{\gamma} \ln \frac{\gamma + (1-\gamma) \cdot x}{x} \quad \text{and} \quad f_\gamma^{-1}(y) = \frac{\gamma \cdot e^{-\gamma \cdot y}}{1 - (1-\gamma) \cdot e^{-\gamma \cdot y}}$$

$$\text{and} \quad H_\gamma^*(x,y) = \frac{x + y - x \cdot y - (1-\gamma)xy}{1 - (1-\gamma) \cdot xy} \tag{3.12}$$

is a *t*-conorm with increasing generator

$$g_\gamma(x) = \frac{1}{\gamma} \ln \frac{\gamma + (1-\gamma) \cdot (1-x)}{1-x}$$

and

$$g_\gamma^{-1}(y) = \frac{1 - e^{-\gamma \cdot y}}{1 - (1-\gamma) \cdot e^{-\gamma \cdot y}}$$

3. Dombi [2] also proposed *t*-norm and *t*-conorm that are algebraic:

$$D(x,y) = \frac{1}{1 + \left(\left(\frac{1}{x} - 1 \right)^\lambda + \left(\frac{1}{y} - 1 \right)^\lambda \right)^{\frac{1}{\lambda}}} \tag{3.13}$$

$$D^*(x,y) = \cfrac{1}{1+\left(\left(\dfrac{1}{x}-1\right)^{-\lambda}+\left(\dfrac{1}{y}-1\right)^{-\lambda}\right)^{-\frac{1}{\lambda}}}$$

4. Frank [7] proposed logarithmic *t*-norm and *t*-conorms:

$$\begin{aligned} F(x,y) &= \log_a\left(1+\frac{(a^x-1)(a^y-1)}{a-1}\right) \\ F^*(x,y) &= 1-\log_a\left(1-\frac{(a^{1-x}-1)(a^{1-y}-1)}{a-1}\right) \end{aligned}, \quad a \in (0,\infty) \qquad (3.14)$$

3.3.3 Negation

A function n: [0, 1] → [0, 1] is called negation iff

1. $n(0) = 1$, $n(1) = 0$
2. n is non-increasing

Negation is called strict iff

1. n is decreasing
2. n is continuous

Strict negation is called involution iff
$n(n(a)) = a$ for all a

Fuzzy *t*-norm and *t*-conorm are used in many areas of medical image processing such as in image enhancement, segmentation and morphology. These are shown later in different chapters.

3.4 Fuzzy Aggregating Operators

In many cases in multi-criteria decision-making problems, the type of aggregation may not be pure 'ANDing' of *t*-norms or pure 'ORing' of *t*-conorms. Yager [19] introduced an aggregation technique based on the ordered weighted averaging (OWA) operator that provides an aggregation lying in the middle of these two extremes (to satisfy all the criteria or at least one of the criteria). A special case of 'mean' operator may be used. There are many operators for aggregation information [2,5,10,11,19], and the two common operators used are the weighted averaging (WA) operator and OWA operator.

3.4.1 Weighted Averaging Operator

A WA operator of dimension 'n' is a mapping WA : $R^n \to R$ that is associated with the 'n' vector of weights: $w = (w_1, w_2, \ldots, w_n)^T$ to the vector $a_i (i = 1, 2, 3, \ldots, n)$ such that

$$\sum_{i=1}^{n} w_i = 1 \quad \text{and} \quad w_i \in [0,1]$$

So, $F(a_1, a_2, \ldots a_n) = \sum_{i=1}^{n} w_i a_i = w_1 a_1 + w_2 a_2 + \cdots + w_n a_n$

The weighted operator first weights all the arguments and then aggregates all these weighted arguments.

3.4.2 Ordered Weighted Averaging Operator

An OWA operator is almost the same as the WA operator, but reordering is given to all arguments. First, it reorders all the arguments in descending order; next, weights are given to these arguments; and finally, the OWA operator aggregates all the ordered arguments.

An OWA operator of dimension 'n' is a mapping OWA : $R^n \to R$ that has an associated set of weights: $w = (w_1, w_2, \ldots, w_n)^T$ to the vector $a_i (i = 1, 2, 3, \ldots, n)$ such that

$$\sum_{i=1}^{n} w_i = 1 \quad \text{and} \quad w_i \in [0,1]$$

So, $F(a_1, a_2, \ldots, a_n) = \sum_{i=1}^{n} w_i a_{\sigma(i)}$

where
$\sigma = 1, 2, 3, \ldots, n$
$a_{\sigma(i)}$ is the ith largest value in the set $(a_1, a_2, \ldots, a_n)$ such that $a_{\sigma(i)} \geq a_{\sigma(i-1)}$

3.5 Intuitionistic Fuzzy Weighted Averaging Operator

The averaging operator, as described in fuzzy sense, is extended using intuitionistic fuzzy set (IFS) where both membership and non-membership functions are taken into account. Intuitionistic fuzzy aggregation operators are studied by many authors [10,17,20]. An IFS in X is given as

$$A = \{X, \mu_A(x), \nu_A(x) | x \in A\},$$

where $\mu_A(x)$ and $\nu_A(x)$ are the membership and non-membership functions, respectively, with the condition $\mu_A(x) + \nu_A(x) + \pi_A(x) = 1$.

The grade of membership is bounded in a subinterval $[\mu_A(x), 1 - \nu_A(x)]$.

Let $a = (\mu_a, \nu_a)$ be the intuitionistic fuzzy value, and $\mu_a \in [0, 1]$, $\nu_a \in (0, 1)$ and $\mu_a + \nu_a \leq 1$. The score functions introduced by Chen and Tan [1] are used to evaluate the degree of suitability in the decision-maker's requirement. It is also called a score function. The score on a is given as

$$S(a) = \mu_a - \nu_a$$

which is used to find the deviation between the membership and non-membership degrees and $S(a) = [-1, 1]$.

Hong and Choi [9] defined the accuracy function H to evaluate the degree of accuracy of the intuitionistic fuzzy value $a = (\mu_a, \nu_a)$ as

$$H(a) = \mu_a + \nu_a$$

where $H(a) \in [0, 1]$.

The larger the value of $H(a)$, the higher the degree of accuracy of the degree of membership of the intuitionistic fuzzy value a. The score function and accuracy function represent the difference and sum of the membership and non-membership values. The score function is used to handle multi-attribute decision-making problems based on intuitionistic fuzzy aggregation operators to aggregate a collection of intuitionistic fuzzy values.

Xu [17,20] defined three laws of intuitionistic fuzzy values that will be used to define the aggregate operator. Consider two IFSs A and B and let the intuitionistic fuzzy values be

$$a_1 = (\mu_{a_1}, \nu_{a_1}) \quad \text{and} \quad a_2 = (\mu_{a_2}, \nu_{a_2}) \text{ then}$$

$$a_1 \oplus a_2 = (\mu_{a_1} + \mu_{a_2} - \mu_{a_1} \cdot \mu_{a_2}, \nu_{a_1} \cdot \nu_{a_2})$$

$$\lambda a = (1 - (1 - \mu_a)^{\lambda}, \nu_a^{\lambda}), \quad \lambda > 0$$

$$a^{\lambda} = (\mu_a^{\lambda}, 1 - (1 - \nu_a)^{\lambda}), \quad \lambda > 0$$

3.5.1 Generalized Intuitionistic Fuzzy Weighted Averaging Operator

Let $a_j = (\mu_{a_j}, \nu_{a_j})$ with $(j = 1, 2, 3, \ldots, n)$ be a collection of intuitionistic fuzzy values; then

$$\text{GIFWA}_w(a_1, a_2, a_3, \ldots, a_n) = \left(w_1 a_1^{\lambda} \oplus w_2 a_2^{\lambda} \oplus w_3 a_3^{\lambda} \oplus \cdots \oplus w_n a_n^{\lambda} \right)^{1/\lambda}$$

$$= \left(\left(1 - \prod_{j=1}^{n} \left(1 - \mu_{a_j}^{\lambda} \right)^{w_j} \right)^{1/\lambda}, 1 - \left(1 - \prod_{j=1}^{n} \left(1 - (1 - \nu_{a_j})^{\lambda} \right)^{w_j} \right)^{1/\lambda} \right)$$

$$(3.15)$$

where $w = (w_1, w_2, \ldots, w_n)^T$ is the weight vector of $a_j(j = 1, 2, 3, \ldots, n)$, and $\sum_{j=1}^{n} w_j = 1$ and $\lambda > 0$. To compute generalized intuitionistic fuzzy weighted averaging (GIFWA), the first two terms of Equation 3.15, that is, for $n = 2$, are computed, that is,

$$w_1 a_1^\lambda \oplus w_2 a_2^\lambda$$

$$a_1^\lambda = \left(\mu_{a_1}^\lambda, 1-(1-v_{a_1})^\lambda\right), \quad a_2^\lambda = \left(\mu_{a_2}^\lambda, 1-(1-v_{a_2})^\lambda\right)$$

Then

$$w_1 a_1^\lambda = \left(1-(1-\mu_{a_1}^\lambda)^{w_1}, \quad (1-(1-v_{a_1})^\lambda)^{w_1}\right)$$

$$w_2 a_2^\lambda = \left(1-\left(1-\mu_{a_2}^\lambda\right)^{w_2}, \quad \left(1-(1-v_{a_2})^\lambda\right)^{w_2}\right)$$

$$w_1 a_1^\lambda \oplus w_2 a_2^\lambda = \left[1-\left(1-\mu_{a_1}^\lambda\right)^{w_1}, \left(1-(1-v_{a_1})^\lambda\right)^{w_1}\right] \oplus \left[1-\left(1-\mu_{a_2}^\lambda\right)^{w_2}, \left(1-(1-v_{a_2})^\lambda\right)^{w_2}\right]$$

$$= \left[\begin{array}{l} 1-\left(1-\mu_{a_1}^\lambda\right)^{w_1}+1-\left(1-\mu_{a_2}^\lambda\right)^{w_2}-\left(1-\left(1-\mu_{a_1}^\lambda\right)^{w_1}\right)\cdot\left(1-\left(1-\mu_{a_2}^\lambda\right)^{w_2}\right), \\ \left(1-(1-v_{a_1})^\lambda\right)^{w_1}\cdot\left(1-(1-v_{a_2})^\lambda\right)^{w_2} \end{array}\right]$$

$$= \left[\begin{array}{l} 2-\left(1-\mu_{a_1}^\lambda\right)^{w_1}-\left(1-\mu_{a_2}^\lambda\right)^{w_2}-1+\left(1-\mu_{a_1}^\lambda\right)^{w_1} \\ +\left(1-\mu_{a_2}^\lambda\right)^{w_2}-\left(1-\mu_{a_1}^\lambda\right)^{w_1}\cdot\left(1-\mu_{a_2}^\lambda\right)^{w_2}, \\ \left(1-(1-v_{a_1})^\lambda\right)^{w_1}\cdot\left(1-(1-v_{a_2})^\lambda\right)^{w_2} \end{array}\right]$$

$$= 1-\left(1-\mu_{a_1}^\lambda\right)^{w_1}\cdot\left(1-\mu_{a_2}^\lambda\right)^{w_2}, \left(1-(1-v_{a_1})^\lambda\right)^{w_1}\cdot\left(1-(1-v_{a_2})^\lambda\right)^{w_2}$$

$$= \left(1-\prod_{j=1}^{2}\left(1-\mu_{a_j}^\lambda\right)^{w_j}, \prod_{j=1}^{2}\left(1-(1-v_{a_j})^\lambda\right)^{w_j}\right)$$

Then by the induction method,

$$w_1 a_1^\lambda \oplus w_2 a_2^\lambda \oplus w_3 a_3^\lambda \oplus \cdots \oplus w_n a_n^\lambda = \left(1-\prod_{j=1}^{n}\left(1-\mu_{a_j}^\lambda\right)^{w_j}, \prod_{j=1}^{n}(1-(1-v_{a_j})^\lambda)^{w_j}\right)$$

$$\text{GIFWA}_w(a_1, a_2, a_3, \dots, a_n) = \left(1 - \prod_{j=1}^{n}\left(1 - \mu_{a_j}^{\lambda}\right)^{w_j}, \prod_{j=1}^{n}\left(1 - (1 - v_{a_j})^{\lambda}\right)^{w_j}\right)^{1/\lambda}$$

$$= \left(\left(1 - \prod_{j=1}^{n}\left(1 - \mu_{a_j}^{\lambda}\right)^{w_j}\right)^{1/\lambda}, 1 - \left(1 - \prod_{j=1}^{n}\left(1 - (1 - v_{a_j})^{\lambda}\right)^{w_j}\right)^{1/\lambda}\right)$$

The following cases hold from the GIFWA operator.

Case 1: If $\lambda = 1$, then GIFWA reduces to IFWA:

$$\text{IFWA}_w(a_1, a_2, a_3, \dots, a_n) = (w_1 a_1 \oplus w_2 a_2 \oplus w_3 a_3 \oplus \dots \oplus w_n a_n)$$

$$= \left(1 - \prod_{j=1}^{n}(1 - \mu_{a_j})^{w_j}, \quad \prod_{j=1}^{n}(1 - (1 - v_{a_j}))^{w_j}\right)$$

$$= \left(1 - \prod_{j=1}^{n}(1 - \mu_{a_j})^{w_j}, \quad \prod_{j=1}^{n} v_{a_j}^{w_j}\right)$$

Case 2: If $w = \left(\dfrac{1}{n}, \dfrac{1}{n}, \dfrac{1}{n}, \dots, \dfrac{1}{n}\right)^T$, then GIFWA reduces to the intuitionistic fuzzy averaging operator: IFA $(a_1, a_2, a_3, \dots, a_n) = (1/n)(a_1 \oplus a_2 \oplus a_3 \oplus \dots \oplus a_n)$

Case 3: If $\lambda \to \infty$, then GIFWA reduces to the intuitionistic fuzzy maximum operator:

$$\text{IFMAX}_w(a_1, a_2, a_3, \dots, a_n) = \max_j(a_j)$$

An example is given to calculate GIFWA of the four intuitionistic fuzz values.

Example 3.1

Let us consider four intuitionistic fuzzy values
$a_1 = (0.1, 0.6)$, $a_2 = (0.4, 0.3)$, $a_3 = (0.6, 0.2)$, $a_4 = (0.2, 0.5)$ with weight vector be $w = (0.2, 0.3, 0.1, 0.4)^T$ of a_j ($j = 1, 2, 3, 4$) and $\lambda = 2$

Solution

From the intuitionistic fuzzy values, we have

$$\mu_{a_1} = 0.1, \quad \mu_{a_2} = 0.4, \quad \mu_{a_3} = 0.6, \quad \mu_{a_4} = 0.2$$

$$v_{a_1} = 0.6, \quad v_{a_2} = 0.3, \quad v_{a_3} = 0.2, \quad v_{a_4} = 0.5$$

Then

$$\text{GIFWA}_w(a_1,a_2,a_3,\ldots,a_n) = \left(\left(1-\prod_{j=1}^{4}\left(1-\mu_{a_j}^\lambda\right)^{w_j}\right)^{1/\lambda}, \; 1-\left(1-\prod_{j=1}^{4}\left(1-(1-v_{a_j})^\lambda\right)^{w_j}\right)^{1/\lambda}\right)$$

$$= \begin{bmatrix} 1-(1-0.1^2)^{0.2} \times (1-0.4^2)^{0.3} \times (1-0.6^2)^{0.1} \times (1-0.2^2)^{0.4})^{1/2}, \\ 1-\left(1-(1-(1-0.6)^2)^{0.2} \times (1-(1-0.3)^2)^{0.3}\right. \\ \left. \times (1-(1-0.2)^2)^{0.1} \times (1-(1-0.5)^2)^{0.4}\right)^{1/2} \end{bmatrix}$$

$$= (0.3330, 0.3959)$$

3.5.2 Generalized Intuitionistic Fuzzy Ordered Weighted Averaging Operator

Similar to the generalized intuitionistic fuzzy ordered weighted averaging (GIFOWA) operator and following a similar type of procedure as the fuzzy ordered weighting operator, let $a_j = (\mu_{a_j}, v_{a_j})$ with ($j = 1, 2, 3, \ldots, n$) be a collection of intuitionistic fuzzy values; then

$$\text{GIFOWA}_w(a_1,a_2,a_3,\ldots,a_n) = \left(w_1 a_{\sigma(1)}^\lambda \oplus w_2 a_{\sigma(2)}^\lambda \oplus w_3 a_{\sigma(3)}^\lambda \oplus \cdots \oplus w_n a_{\sigma(n)}^\lambda\right)^{1/\lambda}$$

$$= \left(\left(1-\prod_{j=1}^{n}\left(1-\mu_{a_{\sigma(j)}}^\lambda\right)^{w_j}\right)^{1/\lambda}, \right.$$

$$\left. 1-\left(1-\prod_{j=1}^{n}\left(1-(1-v_{a_{\sigma(j)}})^\lambda\right)^{w_j}\right)^{1/\lambda}\right) \tag{3.16}$$

where

$w = (w_1, w_2, \ldots, w_n)^T$ is an associated weight vector, and $\sum_{j=1}^{n} w_j = 1$ and $\lambda > 0$
$\sigma = (1, 2, 3, \ldots, n)$, and $a_{\sigma(i)}$ is the jth largest value in the set $(a_1, a_2, \ldots, a_n)$ such that $a_{\sigma(i)} \geq a_{\sigma(i-1)}$.

The following cases hold from the GIFOWA operator:

Case 1: If $\lambda = 1$, then GIFOWA reduces to intuitionistic fuzzy ordered weighted averaging (IFOWA):

$$\text{GIFOWA}_w(a_1, a_2, a_3, \ldots, a_n) = \left(w_1 a_{\sigma(1)} \oplus w_2 a_{\sigma(2)} \oplus w_3 a_{\sigma(3)} \oplus \cdots \oplus w_n a_{\sigma(n)} \right)$$

$$= \left(1 - \prod_{j=1}^{n} \left(1 - \mu_{a_{\sigma(j)}} \right)^{w_j}, 1 - \left(1 - \prod_{j=1}^{n} \left(1 - \left(1 - v_{a_{\sigma(j)}} \right) \right)^{w_j} \right) \right)$$

$$= \left(1 - \prod_{j=1}^{n} \left(1 - \mu_{a_{\sigma(j)}} \right)^{w_j}, \prod_{j=1}^{n} v_{a_{\sigma(j)}}^{w_j} \right)$$

Case 2: If $w = \left(\dfrac{1}{n}, \dfrac{1}{n}, \dfrac{1}{n}, \ldots, \dfrac{1}{n} \right)^T$ and $\lambda = 1$, then GIFOWA reduces to the intuitionistic fuzzy averaging operator:

$$\text{IFA}(a_1, a_2, a_3, \ldots, a_n) = \frac{1}{n} \left(a_1 \oplus a_2 \oplus a_3 \cdots \oplus a_n \right)$$

Case 3: If $\lambda \to \infty$, then GIFWA reduces to the intuitionistic fuzzy maximum operator:

$$\text{IFMAX}_w(a_1, a_2, a_3, \ldots, a_n) = \max_j (a_j)$$

An example is given to calculate GIFOWA of the four intuitionistic fuzzy values.

Example 3.2

Let us consider five intuitionistic fuzzy values
$a_1 = (0.2, 0.6)$, $a_2 = (0.4, 0.3)$, $a_3 = (0.6, 0.2)$, $a_4 = (0.7, 0.1)$, $a_5 = (0.1, 0.7)$ with weight vector be $w = (0.112, 0.236, 0.304, 0.236, 0.112)^T$ of a_j ($j = 1, 2, 3, 4, 5$) and $\lambda = 2$

Solution

From the intuitionistic fuzzy values, we have

$$\mu_{a_1} = 0.2, \quad \mu_{a_2} = 0.4, \quad \mu_{a_3} = 0.6, \quad \mu_{a_4} = 0.7, \quad \mu_{a_5} = 0.1$$

$$v_{a_1} = 0.6, \quad v_{a_2} = 0.3, \quad v_{a_3} = 0.2, \quad v_{a_4} = 0.1, \quad v_{a_5} = 0.7$$

Now, the scores of a_j($j = 1, 2, 3, 4, 5$) are computed as

$$s(a_1) = 0.2 - 0.6 = -0.4, \quad s(a_2) = 0.4 - 0.3 = 0.1, \quad s(a_3) = 0.6 - 0.2 = 0.4,$$

$$s(a_4) = 0.7 - 0.1 = 0.6, \quad s(a_5) = 0.1 - 0.7 = -0.6$$

Since $s(a_4) > s(a_3) > s(a_2) > s(a_1) > s(a_5)$,

$$a_{\sigma(1)} = (0.7, 0.1), \quad a_{\sigma(2)} = (0.6, 0.2), \quad a_{\sigma(3)} = (0.4, 0.3),$$

$$a_{\sigma(4)} = (0.2, 0.6), \quad a_{\sigma(5)} = (0.1, 0.7)$$

Now, from the definition of IFOWA

$$(a_1, a_2, a_3, \ldots, a_n) = \left(1 - \prod_{j=1}^{n} \left(1 - \mu_{a_{\sigma(j)}}\right)^{w_j}, \quad \prod_{j=1}^{n} v_{a_{\sigma(j)}}^{w_j}\right)$$

$$= \begin{bmatrix} \left[1 - (1 - 0.7)^{0.112} \times (1 - 0.6)^{0.236} \times (1 - 0.4)^{0.304}\right. \\ \left. \times (1 - 0.2)^{0.236} \times (1 - 0.1)^{0.112}\right), \\ 0.1^{0.112} \times 0.2^{0.236} \times 0.3^{0.304} \times 0.6^{0.236} \times 0.7^{0.112} \end{bmatrix}$$

$$= (0.4350, 0.3122)$$

3.5.3 Generalized Intuitionistic Fuzzy Hybrid Averaging Operator

The GIFWA operator weighs only intuitionistic fuzzy values, and GIFOWA weighs only the ordered positions of intuitionistic fuzzy values. To overcome the limitation, the generalized intuitionistic fuzzy hybrid averaging (GIFHA) operator is introduced, which weighs both intuitionistic fuzzy values and its ordered position.

A GIFHA operator of dimension 'n' with an associated vector $w = (w_1, w_2, \ldots, w_n)^T$ and $\sum_{j=1}^{n} w_j = 1$, $\dot{a}_{\sigma(j)}$ is the jth largest of the weighted intuitionistic fuzzy values $\dot{a}_j (\dot{a}_j = n\omega_j a_j)$, $j = 1, 2, 3, \ldots, n$ and $\omega = (\omega_1, \omega_2, \omega_3, \ldots, \omega_n)^T$ is the weight vector of $a_j (j = 1, 2, 3, \ldots, n)$ and $\sum_{j=1}^{n} \omega_j = 1$ and $\lambda > 0$ is given as

$$\text{GIFHA}_{w,\omega}(a_1, a_2, a_3, \ldots, a_n) = \left(w_1 \dot{a}_{\sigma(1)}^{\lambda} \oplus w_2 \dot{a}_{\sigma(2)}^{\lambda} \oplus w_3 \dot{a}_{\sigma(3)}^{\lambda} \cdots \oplus w_n \dot{a}_{\sigma(n)}^{\lambda}\right)^{1/\lambda}$$

$$= \left(\left(1 - \prod_{j=1}^{n}\left(1 - \mu_{\dot{a}_{\sigma(j)}}^{\lambda}\right)^{w_j}\right)^{1/\lambda},\right.$$

$$\left.1 - \left(1 - \prod_{j=1}^{n}\left(1 - \left(1 - v_{\dot{a}_{\sigma(j)}}\right)^{\lambda}\right)^{w_j}\right)^{1/\lambda}\right) \tag{3.17}$$

where $\dot{a}_{\sigma(j)} = (\mu_{\dot{a}_{\sigma(j)}}, v_{\dot{a}_{\sigma(j)}})$.

With $\lambda = 1$, GIFHA reduces to the intuitionistic fuzzy hybrid averaging (IFHA) operator:

$$\text{IFHA}_{w,\omega}(a_1,a_2,a_3,\ldots,a_n) = \left(1-\prod_{j=1}^{n}\left(1-\mu_{\dot{a}_{\sigma(j)}}\right)^{w_j}, \prod_{j=1}^{n}v_{\dot{a}_{\sigma(j)}}^{w_j}\right)$$

An example is given to show the computation of IFHA for the five intuitionistic values.

Example 3.3

Let us consider five intuitionistic fuzzy values
$a_1 = (0.2, 0.6)$, $a_2 = (0.4, 0.2)$, $a_3 = (0.5, 0.6)$, $a_4 = (0.5, 0.4)$, $a_5 = (0.7, 0.2)$ and weight vector be $\omega = (0.25, 0.20, 0.15, 0.18, 0.22)^T$ of a_j ($j = 1, 2, 3, 4, 5$) and $\lambda = 2$

Solution
Using the formula $\lambda a = (1 - (1 - \mu_a)^\lambda, v_a^\lambda)$, $\lambda > 0$,
The weighted intuitionistic fuzzy values are obtained as

$$\dot{a}_j = n\omega_j a_j = 5\omega_j a_j$$

$$\dot{a}_1 = [1-(1-0.2)^{5*0.25}, 0.6^{5*0.25}] = [0.243, 0.528]$$

$$\dot{a}_2 = [1-(1-0.4)^{5*0.2}, 0.2^{5*0.2}] = [0.4, 0.2]$$

$$\dot{a}_3 = [1-(1-0.5)^{5*0.15}, 0.6^{5*0.15}] = [0.4054, 0.6817]$$

$$\dot{a}_4 = [1-(1-0.5)^{5*0.18}, 0.4^{5*0.18}] = [0.4641, 0.4384]$$

$$\dot{a}_5 = [1-(1-0.7)^{5*0.22}, 0.2^{5*0.22}] = [0.7340, 0.1703]$$

Now, the scores of $\dot{a}_j (j = 1, 2, 3, 4, 5)$ are computed as

$s(\dot{a}_1) = 0.243 - 0.528 = -0.285$, $\quad s(\dot{a}_2) = 0.4 - 0.2 = 0.2$, $\quad s(\dot{a}_3) = 0.4054 - 0.6817 = -0.2763$,

$s(\dot{a}_4) = 0.4641 - 0.4384 = 0.0257$, $\quad s(\dot{a}_5) = 0.7340 - 0.1703 = 0.5637$

As $s(\dot{a}_5) > s(\dot{a}_2) > s(\dot{a}_4) > s(\dot{a}_3) > s(\dot{a}_1)$,
So,

$$\dot{a}_{\sigma(1)} = (0.7340, 0.1703), \quad \dot{a}_{\sigma(2)} = (0.4, 0.2), \quad \dot{a}_{\sigma(3)} = (0.4641, 0.4384),$$
$$\dot{a}_{\sigma(4)} = (0.4054, 0.6817), \quad \dot{a}_{\sigma(5)} = (0.243, 0.528)$$

Then

$$\mu_{\dot{a}\sigma(1)} = 0.734, \quad \mu_{\dot{a}\sigma(2)} = 0.4, \quad \mu_{\dot{a}\sigma(3)} = 0.464, \quad \mu_{\dot{a}\sigma(4)} = 0.405, \quad \mu_{\dot{a}\sigma(5)} = 0.243$$

$$v_{\dot{a}\sigma(1)} = 0.170, \quad v_{\dot{a}\sigma(2)} = 0.2, \quad v_{\dot{a}\sigma(3)} = 0.438, \quad v_{\dot{a}\sigma(4)} = 0.682, \quad v_{\dot{a}\sigma(5)} = 0.528$$

Suppose $w = [0.112, 0.236, 0.304, 0.236, 0.112]^T$ which is derived from the normal distribution–based method [16]. A brief idea on the normal distribution–based method is given:

Let $w = (w_1, w_2, \ldots, w_n)^T$ be a weight vector $(i = 1, 2, 3, \ldots, n)$, and μ_n = mean of the collection which is computed as

$$\mu_n = \frac{1}{n}\frac{n(n+1)}{2} = \frac{n+1}{2}$$

σ_n is the standard deviation $(\sigma > 0)$ which is computed as

$$\sigma_n = \sqrt{\frac{1}{n}\sum_{i=1}^{n}(i-\mu_n)^2}$$

With μ_n and σ_n, w_i becomes

$$w_i = \frac{\dfrac{1}{\sqrt{2\pi}\sigma_n}e^{-\left[\frac{(i-\mu_n)^2}{2\sigma_n^2}\right]}}{\sum_{j=1}^{n}\dfrac{1}{\sqrt{2\pi}\sigma_n}e^{-\left[\frac{(j-\mu_n)^2}{2\sigma_n^2}\right]}} = \frac{e^{-\left[\frac{(i-\mu_n)^2}{2\sigma_n^2}\right]}}{\sum_{j=1}^{n}e^{-\left[\frac{(j-\mu_n)^2}{2\sigma_n^2}\right]}}$$

With $\mu_n = (n+1)/2$ and $\sigma_n = \sqrt{(1/n)\sum_{i=1}^{n}(i-\mu_n)^2}$

$$w_i = \frac{e^{-\left[\frac{(i-\mu_n)^2}{2\sigma_n^2}\right]}}{\sum_{j=1}^{n}e^{-\left[\frac{(j-\mu_n)^2}{2\sigma_n^2}\right]}} = \frac{e^{-\left[\left(i-\frac{n+1}{2}\right)^2\right]/2\sigma_n^2}}{\sum_{j=1}^{n}e^{-\left[\left(j-\frac{n+1}{2}\right)^2\right]/2\sigma_n^2}}, \quad i = 1, 2, 3, \ldots, n$$

Now, for the earlier problem, we get

$$\mu_n = \frac{5+1}{2} = 3 \quad \text{and} \quad \sigma_n = \sqrt{\frac{1}{n}\sum_{i=1}^{n}(i-\mu_n)^2}$$

$$= \sqrt{\frac{1}{5}\left((1-3)^2 + (2-3)^2 + (3-3)^2 + (4-3)^2 + (5-3)^2\right)} = \sqrt{2}$$

So,

$$w_i = \frac{e^{-[(i-3)^2]/4}}{\sum_{j=1}^{n} e^{-[(j-3)^2]/4}}$$

$$w_1 = \frac{e^{-[(1-3)^2/4]}}{e^{-[(1-3)^2/4]} + e^{-[(2-3)^2/4]} + e^{-[(3-3)^2/4]} + e^{-[(4-3)^2/4]} + e^{-[(5-3)^2/4]}}$$

$$= \frac{e^{-1}}{e^{-1} + e^{-1/4} + e^{-0} + e^{-1/4} + e^{-1}} = \frac{0.3678}{3.2932} = 0.1117 \approx 0.112$$

$$w_2 = \frac{e^{-[(2-3)^2/4]}}{e^{-[(1-3)^2/4]} + e^{-[(2-3)^2/4]} + e^{-[(3-3)^2/4]} + e^{-[(4-3)^2/4]} + e^{-[(5-3)^2/4]}} = \frac{0.7788}{3.2932} = 0.236$$

$$w_3 = \frac{e^{-[(3-3)^2/4]}}{e^{-[(1-3)^2/4]} + e^{-[(2-3)^2/4]} + e^{-[(3-3)^2/4]} + e^{-[(4-3)^2/4]} + e^{-[(5-3)^2/4]}} = \frac{1}{3.2932} = 0.304$$

In a similar way, other weight values, $w_4 = 0.236$, $w_5 = 0.112$ with $i = 2, 3, 4, 5$, are obtained.

Now, with the weight vector,

$$\text{IFHA}_w(a_1, a_2, a_3, \ldots, a_n) = \left(1 - \prod_{j=1}^{5} \left(1 - \mu_{\dot{a}_{\sigma(j)}}\right)^{w_j}, \prod_{j=1}^{5} v_{\dot{a}_{\sigma(j)}}^{w_j}\right)$$

$$= \begin{bmatrix} (1 - (1 - 0.734)^{0.112} \cdot (1 - 0.4)^{0.236} \cdot (1 - 0.464)^{0.304} \\ \cdot (1 - 0.405)^{0.236} \cdot (1 - 0.234)^{0.112}), 0.170^{0.112} \cdot 0.2^{0.236} \\ \cdot 0.438^{0.304} \cdot 0.682^{0.236} \cdot 0.528^{0.112} \end{bmatrix}$$

So, $\text{IFHA}_w(a_1, a_2, a_3, \ldots, a_n) = [0.4571, 0.3712]$.

3.6 Application of Intuitionistic Fuzzy Operators to Multi-Attribute Decision-Making

Let $A = A_1, A_2, A_3, \ldots, A_m$ and $B = B_1, B_2, B_3, \ldots, B_n$ be a set of attributes and $\omega = (\omega_1, \omega_2, \omega_3, \ldots, \omega_n)^T$ be the weight vector of $B_j (j = 1, 2, 3, \ldots, n)$ with $\sum_{j=1}^{n} \omega_j = 1$. The characteristic information of all the 'm' alternatives is represented by an IFS:

$$A_i = \{(B_1, (\mu_{i1}, v_{i1})), (B_2, (\mu_{i2}, v_{i2})), (B_3, (\mu_{i3}, v_{i3})), \ldots, (B_n, (\mu_{in}, v_{in}))\}$$

$$= \{B_j, (\mu_{ij}, v_{ij})\}$$

The characteristic information of all the alternatives is represented by intuitionistic fuzzy values $a_{ij}(i = 1, 2, 3, \ldots, m; j = 1, 2, 3, \ldots, n)$. If $a_{ij} = (\mu_{ij}, \nu_{ij})$, then $A_i = \{\langle B_j, a_{ij}\rangle, B_j \in B\}$ where μ_{ij} denotes the degree to which the alternative A_i satisfies the attribute B_j and ν_{ij} denotes the degree to which the alternative A_i does not satisfy the attribute B_j with $\mu_{ij} \in [0, 1]$, $\nu_{ij} \in [0, 1]$, respectively.

To get the best alternative, one can use GIFWA, GIFOWA and GIFHA to derive $a_i a_i = (\mu_{ai}, \nu_{ai})$ of the alternatives A_i as

$$a_i = \text{GIFWA}_w(a_{i1}, a_{i2}, a_{i3}, \ldots, a_{in}), \quad i = 1, 2, 3, \ldots, m$$

$$a_i = \text{GIFOWA}_w(a_{i1}, a_{i2}, a_{i3}, \ldots, a_{in}), \quad i = 1, 2, 3, \ldots, m$$

$$a_i = \text{GIFHA}_w(a_{i1}, a_{i2}, a_{i3}, \ldots, a_{in}), \quad i = 1, 2, 3, \ldots, m$$

with $w = (w_1, w_2, \ldots, w_n)^T$ as the weight vector related to the operators with $\sum_{j=1}^{n} w_j = 1$, which can be derived from the normal distribution–based method.

Scores $s(a_i)$ ($i = 1, 2, 3, \ldots, m$) of the overall a_i ($i = 1, 2, 3, \ldots, m$) are calculated. Then the alternatives are ranked according to the score values, and the best one is selected.

3.7 Intuitionistic Fuzzy Triangular Norms and Triangular Conorms

Intuitionistic fuzzy t-norms and t-conorms can be constructed using t-norms and t-conorms [4]. IFSs can also be visualized as L-fuzzy sets according to Deschrijver and Kerre [3]. If L^* is a set and $\leq_{L^*}$ is the operation,

$$L^* = \left\{(a_1, a_2) \mid (a_1, a_2) \in [0, 1]^2 \wedge a_1 + a_2 \leq 1\right\}$$

$$(a_1, a_2) \leq_{L^*} (b_1, b_2) \Leftrightarrow a_1 \leq b_1 \wedge a_2 \geq b_2 \quad \text{for all } (a_1, a_2), (b_1, b_2) \in L^*$$

Then $(L^*, \leq_{L^*})$ is a complete lattice.

Using the lattice $(L^*, \leq_{L^*})$, triangular norms and conorms can be extended to intuitionistic fuzzy case.

An intuitionistic fuzzy triangular norm is a binary operation $\mathfrak{I}$ which is increasing, commutative, associative and satisfying, $\mathfrak{I}(x, 1_{L^*}) = x$. It is a mapping $\mathfrak{I} : (L^*)^2 \to L^*$ such that

1. Commutativity: $\mathfrak{I}(x, y) = \mathfrak{I}(y, x)$
2. Associativity: $\mathfrak{I}(x, \mathfrak{I}(y, z)) = \mathfrak{I}(\mathfrak{I}(x, y), z)$

3. $\mathfrak{I}(x, z) = \mathfrak{I}(y, z),\ x \leq_{L^*} y$

4. Border condition: $\mathfrak{I}(x, (1, 0)) = x$

An intuitionistic fuzzy triangular conorm is a binary operation S which is increasing, commutative, associative and satisfying, $S(x, 0_{L^*}) = x$. It is a mapping such that

1. Commutativity: $S(x, y) = S(y, x)$
2. Associativity: $S(x, S(y, z)) = S(S(x, y), z)$
3. $S(x, z) = S(y, z), x \leq_{L^*} y$
4. Border condition: $S(x, (0, 1)) = x$

In the earlier formulation, $1_{L^*} = (1, 0),\ 0_{L^*},\ (0, 1)$.

As in fuzzy set theory, fuzzy t-norm and t-conorm are modelled as union and intersection, and intuitionistic fuzzy t-norm and t-conorm can be modelled as union and intersection, respectively.

Let T be the t-norm and S be the t-conorm. If $\mathfrak{I}(x, y) \leq 1 - S(1 - x, 1 - y)$ for all $x, y \in [0, 1]$, then

$$\mathfrak{I}(x, y) = (T(x_1, y_1), S(x_2, y_2))\ \text{is an intuitionistic fuzzy } t\text{-norm}$$

$$S(x, y) = (S(x_1, y_1), T(x_2, y_2))\ \text{is an intuitionistic fuzzy } t\text{-conorm}$$

For two elements (x_1, x_2) and (y_1, y_2), some examples of intuitionistic fuzzy t-norms are

1. $\mathfrak{I}(x, y) = (\max(0, x_1 + y_1 - 1), \min(1, x_2 + y_2))$
2. $\mathfrak{I}(x, y) = (x_1 y_1, x_2 + y_2 - x_2 y_2)$
3. $\mathfrak{I}(x, y) = (\min(x_1, y_1), \max(x_2, y_2))$

3.8 Summary

This chapter summarizes various types of operators such as fuzzy aggregation operators, namely, WA and OWA operators, and fuzzy triangular operators, namely, t-norms and t-conorms. Different types of t-norms and t-conorms are also discussed. Extension of fuzzy aggregation operators to intuitionistic fuzzy case such as intuitionistic fuzzy aggregation operators, namely, the IFWA, IFOWA and IFHA operators, is discussed with examples. Application of intuitionistic fuzzy operator in multi-attribute decision-making is also provided as well as extension of t-norms and t-conorms to IFSs to intuitionistic fuzzy triangular t-norms and t-conorms.

These operators are required in various image-processing applications such as enhancement, segmentation and morphology, which will be shown in subsequent chapters.

References

1. Chen, S.M. and Tan, J.M., Handling multicriteria fuzzy decision-making problems based on vague set theory, *Fuzzy Sets and Systems*, 67, 163–172, 1994.

2. Dombi, J., A general class of fuzzy operators A De Morgan's class of fuzzy operators and fuzziness induced by fuzzy operators, *Fuzzy Sets and Systems*, 8, 149–163, 1982.

3. Deschrijver, G. and Kerre, E.E., On the relationship between some extensions of fuzzy set theory, *Fuzzy Sets and Systems*, 12(1), 45–61, 2004.

4. Deschrijver, G. and Kerre, E.E., On the representation of intuitionistic fuzzy t norms and t co norms, *IEEE Transaction on Fuzzy Systems*, 12(1), 45–61, 2004.

5. Dubois, D. and Prade, H., New results about properties and semantics of fuzzy set theoretic operators, in *Fuzzy Sets: Theory and Applications to Policy Analysis and Information Systems*, P. Wang and S. Chang (Eds.), Plenum Press, New York, pp. 59–65, 1980.

6. Frank, M.J., On simultaneous associativity of $F(x,y)$ and $x+y-F(x,y)$, *Aequationes Mathematicae*, 19, 194–226, 1979.

7. Hamacher, H., *Über logische Aggregation nicht-binär explizierter Entscheidnungskriterien*, R.G. Fisher Verlag, Frankfurt, Germany, 1978.

8. Harsanyi, J.C., Cardinal welfare, individualistic ethics and interpersonal comparisons of utility, *Journal of Political Economics*, 63, 309–321, 1955.

9. Hong, D.H. and Choi, C.H., Multicriteria fuzzy decision-making problems based on vague set theory, *Fuzzy Sets and Systems*, 114, 103–113, 2000.

10. Liu, P., Some Hamacher aggregation operators based on the interval valued intuitionistic fuzzy numbers and their application to decision making, *IEEE Transaction on Fuzzy Systems*, 2(1), 83–97, 2014.

11. Mitchell, H.B., An intuitionistic OWA operator, *International Journal of Uncertainty, Fuzziness and Knowledge-Based Systems*, 12(6), 843–860, 2004.

12. Roychowdhury, S. and Wang, B.H., Composite generalization of Dombi class and a new family of T operators using additive-product connective generator, *Fuzzy Sets and Systems*, 66, 329–346, 1994.

13. Shweizer, B. and Sklar, A., Associative functions and statistical triangle inequalities, *Publicationese Mathamaticae Debrecen*, 10, 69–81, 1963.

14. Sugeno, M., Fuzzy measures and fuzzy integrals: A survey, in *Fuzzy Automata and Decision Process*, M. Gupta, G.N. Saridis and B.R. Gaines (Eds.), North Holland, Amsterdam, the Netherlands, pp. 82–10, 1977.

15. Weber, S., A general concept of fuzzy connectives, negations and implications based on t norms and t Co norms, *Fuzzy Sets and Systems*, 11, 115–134, 1983.

16. Xu, X.S., An overview of methods for determining OWA weights, *International Journal of Intelligent Systems*, 20, 843–865, 2005.
17. Xu, Z., Intuitionistic fuzzy aggregation operators, *IEEE Transaction on Fuzzy Systems*, 15(6), 1179–1187, 2007.
18. Yager, R.R., On the general class of fuzzy connectives, *Fuzzy Sets and Systems*, 4, 235–242, 1980.
19. Yager, R.R., On ordered weighted averaging aggregation operators in multi-criteria decision making, *IEEE Transactions on Systems, Man and Cybernetics*, 18, 183–190, 1988.
20. Zhao, H. et al., Generalized aggregation operators for intuitionistic fuzzy sets, *Journal of Intelligent and Fuzzy Systems*, 25, 1–30, 2010.

4

Similarity, Distance Measures and Entropy

4.1 Introduction

Most of the problems in medical science, engineering and environmental science do not always involve crisp data. And traditional methods may not be used successfully when uncertainties are involved. Many new approaches and theories are introduced since the introduction of fuzzy set and have showed successful applications in various fields. These are similarity, distance or entropy measures that are the vital in image processing and are used in many image processing applications such as image retrieval, registration and segmentation. Similarity measure indicates that the degree of similarity between two fuzzy sets and entropy denotes the fuzziness in a fuzzy set. But fuzzy sets consider only one uncertainty which is the degree of belongingness or membership degree. But in reality, it may not always be certain that the non-membership degree in a fuzzy set is just equal to 1 minus the degree of membership. Many researchers extended fuzzy measures [1,7,10,13,24] using intuitionistic fuzzy set (*IFS*) which are characterized by membership and non-membership functions. Intuitionistic fuzzy–based models may be adequate in many situations when we face human opinions such as 'yes' or 'no' or 'does not apply', more specifically in voting where people can vote for or against or does not vote. Using these sets, new approaches such as fuzzy distance/similarity/entropy measures are extended. Such a generalization of fuzzy set gives us additional information to represent imperfect knowledge that will help in describing many real-time problems accurately. Different types of similarity/distance/ entropy measures using *IFS* are discussed later.

4.2 Similarity Measure

4.2.1 Similarity/Distance Measure

A function $S: F(X)^2 \rightarrow [0, \infty]$ is called a similarity measure between two *IFS*s A and B if it satisfies the following properties:

1. $S(A, B) = S(B, A)$.
2. For three *IFS* sets $A, B, C \, \forall \, A, B, C \in F(X)$, if $A \subseteq B \subseteq C$, then $S(A, B) \geq S(A, C)$ and $S(B, C) \geq S(A, C)$.
3. $S(A, B) = 1$ if $A = B$.
4. $0 \leq S(A, B) \leq 1 \, \forall \, D \in F(X)$.

When the distance is small, nothing can be said about the similarity based on pure distance, when the complement of the object is not taken into account. So, in some situations, pure distance is not a proper measure of similarity. IFSs take into account the non-membership degree, which is used when the distance between the two objects is small, but actually the objects are not similar.

We know that the distance measure and similarity measure are dual concepts. Therefore, one may use the distance measure than the similarity measure [8]. If f is a monotonically decreasing function and since $0 \leq d(A, B) \leq 1$, $f(1) \leq f(d(A, B)) \leq f(0)$ may be written as

$$0 \leq \frac{f(d(A,B)) - f(1)}{f(0) - f(1)} \leq 1$$

Thus, the similarity measure between A and B is given as

$$S(A,B) = \frac{f(d(A,B)) - f(1)}{f(0) - f(1)}$$

So, f should be defined to obtain a reasonable similarity measure. The simplest form of expressing f is

$$f(x) = 1 - x \text{ and in that case}$$

$$S(A, B) = 1 - d(A, B) \text{ as } f(0) = 1, f(1) = 0$$

Sometimes, exponential operation is very useful in finding the similarity measure:

$$\text{If } f(x) = e^{-x}, \text{ then}$$

$$S(A,B) = \frac{e^{-d(A,B)} - e^{-1}}{1 - e^{-1}}, \quad \text{as } f(0) = 1, f(1) = e^{-1}$$

For a logarithmic function $f(x) = \ln(1 + x)$, the similarity measure is

$$S(A,B) = \frac{\ln(1 + d(A,B)) - \ln(2)}{-\ln(2)} \qquad f(0) = \ln(1) = 0, f(1) = \ln(2)$$
$$= \frac{\ln(2) - \ln(1 + d(A,B))}{\ln(2)},$$

One may choose the inverse function, $f(x) = 1/(1 + x)$; then the similarity relation between two *IFS*s A and B is given as:

$$S(A,B) = \frac{\dfrac{1}{1 + d(A,B)} - \dfrac{1}{2}}{1 - \dfrac{1}{2}} = \frac{1 - d(A,B)}{1 + d(A,B)}$$

4.2.2 Distance Measures

In many practical and theoretical problems, there is a need for many reasons to find the difference between two objects and in that case, the knowledge of distance between two *IFS*s is necessary. Consider two *IFS*s A and B that take into account the membership degree μ, the non-membership degree v and the hesitation degree (or intuitionistic fuzzy index) π in $X = \{x_1, x_2, \ldots, x_n\}$.

A function $D: F(X)^2 \to [0, \infty]$ is called a distance measure between two *IFS*s A and B if it satisfies the following properties:

1. $D(A, B) = D(B, A)$.
2. For three *IFS* sets $A, B, C \forall A, B, C \in F(X)$, if $A \subseteq B \subseteq C$, then $D(A, B) \leq D(A, C)$ and $D(B, C) \leq D(A, C)$.
3. $A = B$ if and only if $D(A, A) = 0$.
4. $0 \leq D(A, B) \leq 1 \forall C \in P(X)$.

___

4.3 Different Types of Distance and Similarity Measures

1. Szmidt and Kacprzyk [18] introduced some popular distance measures between two *IFSs*.

    Let A and B be two *IFSs* where $A = \{(x, \mu_A(x), \nu_A(x)) | x \in X\}$ and $B = \{(x, \mu_B(x), \nu_B(x)) | x \in X\}$, and $\mu_A(x)$ and $\nu_A(x)$ are the membership and non-membership functions, respectively.

    Intuitionistic Euclidean Distance

$$D_{IFS}(A,B)$$

$$= \sqrt{\left(\sum_{i=1}^{n} \left[(\mu_A(x_i) - \mu_B(x_i))^2 + (\nu_A(x_i) - \nu_B(x_i))^2 + (\pi_A(x_i) - \pi_B(x_i))^2 \right] \right)}$$

Intuitionistic Normalized Euclidean Distance

$$D_{IFS}(A,B)$$

$$= \sqrt{\frac{1}{n} \left(\sum_{i=1}^{n} \left[(\mu_A(x_i) - \mu_B(x_i))^2 + (\nu_A(x_i) - \nu_B(x_i))^2 + (\pi_A(x_i) - \pi_B(x_i))^2 \right] \right)}$$

Intuitionistic Hamming Distance

$$D_{IFS}(A,B) = \sum_{i=1}^{n} \left(|\mu_A(x_i) - \mu_B(x_i)| + |\nu_A(x_i) - \nu_B(x_i)| + |\pi_A(x_i) - \pi_B(x_i)| \right)$$

2. Distance measure based on the Hausdorff metric

    Grzegorzewski [6] introduced the distances based on the Hausdorff metric.

    The Hausdorff distance on any two non-empty subsets A and B of a compact metric space is defined as [8]

$$d(A,B) = \max \left\{ \sup_{a \in A} \inf_{b \in B} |a - b|, \sup_{b \in B} \inf_{a \in A} |a - b| \right\} \tag{4.1}$$

Now, if the intervals of sets A and B are $A = [a_1, a_2]$ and $B = [b_1, b_2]$ with $a_1 < a_2$ and $b_1 < b_2$, then the Hausdorff metric becomes

$$d(A,B) = \max \left\{ \left[|a_1 - b_1|, |a_2 - b_2| \right] \right\} \tag{4.2}$$

This can be extended to intuitionistic fuzzy case: if A and B are two IFSs, and due to the hesitation in defining the membership function, the expert is hesitant to the extent $\pi_A(x) = 1 - \mu_A(x) - v_A(x)$, then the intervals of the membership degree for A and B are

$$[\mu_A(x), \mu_A(x) + \pi_A(x)] = [\mu_A(x), 1 - v_A(x)]$$

$$[\mu_B(x), \mu_B(x) + \pi_B(x)] = [\mu_B(x), 1 - v_B(x)]$$

respectively.

Using the Hausdorff metric, the distance may be written as

$$d(A, B) = \max\left\{|\mu_A(x) - \mu_B(x)|, |1 - v_A(x) - (1 - v_B(x))|\right\} \qquad (4.3)$$

3. A different type of distance measure suggested by Wang and Xin [21] is given as

$$d_{IFS}(A, B) = \frac{1}{n}\sum_{i=1}^{n}\left[\begin{array}{c} \dfrac{|\mu_A(x_i) - \mu_B(x_i)| + |v_A(x_i) - v_B(x_i)|}{4} \\[2mm] + \dfrac{\max\left(|\mu_A(x_i) - \mu_B(x_i)|, |v_A(x_i) - v_B(x_i)|\right)}{2} \end{array} \right] \qquad (4.4)$$

4. Song and Zhou [16] suggested some modifications on Wang's distance measure. In Wang's measure, the weights of $|\mu_A(x_i) - \mu_B(x_i)|$, $|v_A(x_i) - v_B(x_i)|$, $\max(|\mu_A(x_i) - \mu_B(x_i)|$ and $|v_A(x_i) - v_B(x_i)|)$ are fixed as $(1/4)$, $(1/2)$. But in practice, the weights are not fixed, so Song introduced a weight different from the distance measure

$$d_{IFS}(A, B) = \frac{1}{n}\sum_{i=1}^{n}\left[\begin{array}{c} \alpha|\mu_A(x_i) - \mu_B(x_i)| + \beta|v_A(x_i) - v_B(x_i)| \\[2mm] + \gamma \cdot \max\left(|\mu_A(x_i) - \mu_B(x_i)|, |v_A(x_i) - v_B(x_i)|\right) \end{array} \right] \qquad (4.5)$$

with $\alpha + \beta + \gamma = 1$ and $\alpha, \beta, \gamma \in [0, 1]$.

But in Equation 4.5, element x_i in set $X = \{x_1, x_2, x_3, ..., x_n\}$ has the same weight parameters α, β and γ. In some cases, there may be some elements that affect the distance between the IFSs. In that case, a weight $w_i > 0, i \in \{1, 2, 3, ..., n\}$ is introduced. Then the distance measure becomes

$$d_{IFS}(A, B)$$

$$= \frac{1}{n}\sum_{i=1}^{n}\frac{w_i}{\sum_{i=1}^{n}w_i}\left[\begin{array}{c} \alpha|\mu_A(x_i) - \mu_B(x_i)| + \beta|v_A(x_i) - v_B(x_i)| \\[2mm] + \gamma \cdot \max\left(|\mu_A(x_i) - \mu_B(x_i)|, |v_A(x_i) - v_B(x_i)|\right) \end{array} \right] \qquad (4.6)$$

5. Similarity using the Hausdorff distance

The Hausdorff distance [14] is a measure of how much two non-empty sets (closed and bounded) A and B in a metric space S resemble each other with respect to their positions. If $d(A, B)$ is a metric, one-way Hausdorff measure is defined as

$$H^*(A,B) = \max_{a \in A} d(a,B), \quad a \in A$$

The Hausdorff distance between two sets A and B is

$$H(A,B) = \max\left\{H^*(A,B), H^*(B,A)\right\}$$

To define the distance measure between two *IFSs* based on the Hausdorff distance, consider two *IFSs* A and B in $X = \{x_1, x_2, x_3, \ldots, x_n\}$ and $I_A(x_i)$ and $I_B(x_i)$ are the subintervals on $[0, 1]$ which is denoted as

$$\begin{aligned} I_A(x_i) &= [\mu_A(x_i), 1 - v_A(x_i)] \\ I_B(x_i) &= [\mu_B(x_i), 1 - v_B(x_i)] \end{aligned}, \quad i = 1,2,3,\ldots,n$$

Let $H(I_A(x_i), I_B(x_i))$ be the Hausdorff distance between $I_A(x_i)$ and $I_B(x_i)$. Then the distance $d(A, B)$ is defined [8] as

$$d(A,B) = \frac{1}{n} \sum_{i=1}^{n} H(I_A(x_i), I_B(x_i)) \tag{4.7}$$

From Equation 4.2, the Hausdorff metric between two intervals $A = [a_1, a_2]$ and $B = [b_1, b_2]$ in sets A and B is

$$d(A,B) = \max\left\{\left[|a_1 - b_1|, |a_2 - b_2|\right]\right\}$$

So, Equation 4.7 is written as

$$d(A,B) = \frac{1}{n} \sum_{i=1}^{n} \max\left(|\mu_A(x_i) - \mu_B(x_i)|, |v_B(x_i) - v_A(x_i)|\right)$$

And the similarity measure is $S_H(A, B) = 1 - d_H(A, B)$.

6. Dengfeng and Chuntian [5] suggested another similarity measure

$$S_{IFS}(A,B) = 1 - \frac{1}{\sqrt[p]{n}} \sqrt[p]{\sum_{i=1}^{n} |\varphi_A(x_i) - \varphi_B(x_i)|^p} \tag{4.8}$$

where $1 \leq p < +\infty$ and

$$\varphi_A(x_i) = \frac{\mu_A(x_i) + 1 - \nu_A(x_i)}{2}, \quad \varphi_B(x_i) = \frac{\mu_B(x_i) + 1 - \nu_B(x_i)}{2}$$

7. Hung and Wang [9] suggested some similarity measures that are the extensions of fuzzy sets which are given as

$$S(A, B) = 1 - \frac{1}{2}\left(\max_i |\mu_A(x_i) - \mu_B(x_i)| + \max_i |\nu_A(x_i) - \nu_B(x_i)| \right) \qquad (4.9)$$

$$S(A, B) = 1 - \frac{\sum_{i=1}^{n} |\mu_A(x_i) - \mu_B(x_i)| + \|\nu_A(x_i) - \nu_B(x_i)\|}{\sum_{i=1}^{n} |\mu_A(x_i) + \mu_B(x_i)| + \|\nu_A(x_i) + \nu_B(x_i)\|} \qquad (4.10)$$

$$S(A, B) = \frac{1}{N} \frac{\sum_{i=1}^{n} \min(\mu_A(x_i), \mu_B(x_i)) + \min(\nu_A(x_i), \nu_B(x_i))}{\sum_{i=1}^{n} \max(\mu_A(x_i), \mu_B(x_i)) + \max(\nu_A(x_i), \nu_B(x_i))} \qquad (4.11)$$

8. Liang and Shi [12] suggested another similarity measure between two *IFS*s with intervals $[\mu_A(x_i), 1 - \nu_A(x_i)]$ and $[\mu_B(x_i), 1 - \nu_B(x_i)]$ as

$$S_{IFS}^{L}(A, B) = 1 - \frac{1}{\sqrt[p]{n}} \sqrt[p]{\sum_{i=1}^{n} (\varphi_A(x_i) + \varphi_B(x_i))^p} \qquad (4.12)$$

where $1 \leq p < +\infty$ and

$$\varphi_A(x_i) = \frac{|\mu_A(x_i) - \mu_B(x_i)|}{2}, \quad \varphi_B(x_i) = \left| \frac{1 - \nu_A(x_i)}{2} - \frac{1 - \nu_B(x_i)}{2} \right|$$

9. Liang and Shi [12] suggested another similarity measure. For an *IFS* A with intervals $[\mu_A(x_i), 1 - \nu_A(x_i)]$, the median value of the interval is

$$m_A(x_i) = \frac{\mu_A(x_i) + (1 - \nu_A(x_i))}{2}$$

With this median value, the interval is divided into two subintervals which are denoted as

$$[\mu_A(x_i), m_A(x_i)] \quad \text{and} \quad [m_A(x_i), 1 - \nu_A(x_i)]$$

The median values of two subintervals are written as

$$m_{A1}(x_i) = \frac{\mu_A(x_i) + m_A(x_i)}{2} \quad \text{and} \quad m_{A2}(x_i) = \frac{m_A(x_i) + (1 - v_A(x_i))}{2}$$

In a similar way, the median value of the interval of set B, $[\mu_B(x_i), 1 - v_B(x_i)]$ is

$$m_B(x_i) = \frac{\mu_B(x_i) + (1 - v_B(x_i))}{2}$$

and the two subintervals are $[\mu_B(x_i), m_B(x_i)]$ and $[m_B(x_i), 1 - v_B(x_i)]$.
Likewise, the median values of the two subintervals are

$$m_{B1}(x_i) = \frac{\mu_B(x_i) + m_B(x_i)}{2} \quad \text{and} \quad m_{B2}(x_i) = \frac{m_B(x_i) + (1 - v_B(x_i))}{2}$$

Then the similarity measure is defined as

$$S_{IFS}^L(A, B) = 1 - \frac{1}{\sqrt[p]{n}} \sqrt[p]{\sum_{i=1}^{n} (\varphi_A(x_i) + \varphi_B(x_i))^p} \tag{4.13}$$

where
$$\varphi_A(x_i) = \frac{|m_{A1}(x_i) - m_{B1}(x_i)|}{2}$$

$$\varphi_B(x_i) = \frac{|m_{A2}(x_i) - m_{B2}(x_i)|}{2}$$

4.4 Intuitionistic Fuzzy Measure

A distance measure called *intuitionistic fuzzy divergence* (IFD) is described in [3], where three parameters, namely, the membership degree, the non-membership degree, and the hesitation degree (or intuitionistic fuzzy index), are considered.

Let $A = \{(x, \mu_A(x), v_A(x)) | x \in X\}$ and $B = \{(x, \mu_B(x), v_B(x)) | x \in X\}$ be two IFSs. Considering the hesitation degree, the interval or range of the membership degree of the two IFSs A and B may be represented as $\{\mu_A(x), (\mu_A(x) + \pi_A(x))\}$ and $\{\mu_B(x), (\mu_B(x) + \pi_B(x))\}$ where $\mu_A(x)$ and $\mu_B(x)$ are the membership degrees and $\pi_A(x)$ and $\pi_B(x)$ are the hesitation degrees, with $\pi_A(x) = 1 - \mu_A(x) - v_A(x)$ and $\pi_B(x) = 1 - \mu_B(x) - v_B(x)$. The interval is due to the hesitation or the lack of knowledge in assigning the membership values.

In an image of size $M \times M$ with L distinct grey levels having probabilities $p_0, p_1, ..., p_{L-1}$, the exponential entropy is defined as

$$H = \sum_{i=0}^{L-1} p_i e^{1-p_i}$$

In fuzzy cases, the fuzzy entropy of an image A of size $M \times M$ is defined as

$$H(A) = \frac{1}{n\left(\sqrt{e}-1\right)} \sum_{i=0}^{M-1}\sum_{j=0}^{M-1}\left[\mu_A(a_{ij}) \cdot e^{1-\mu_A(a_{ij})} + (1-\mu_A(a_{ij})) \cdot e^{\mu_A(a_{ij})} - 1\right]$$

where
$n = M^2$
$i, j = 0, 1, 2, ..., M - 1$
$\mu_A(a_{ij})$ is the membership degree of the (i, j)th pixel a_{ij} in the image A

For two images A and B, at the (i, j)th pixels (i.e. at pixels a_{ij} and b_{ij}), the amount of information between the membership degrees of images A and B is given as follows:

1. Due to $m_1(A)$ and $m_1(B)$, that is, $\mu_A(a_{ij})$ and $\mu_B(b_{ij})$ of the (i, j)th pixels

$$\frac{e^{\mu_A(a_{ij})}}{e^{\mu_B(b_{ij})}} \quad \text{or} \quad e^{\mu_A(a_{ij})-\mu_B(b_{ij})}$$

2. Due to $m_2(A)$ and $m_2(B)$, that is, $\mu_A(a_{ij}) + \pi_A(b_{ij})$ and $\mu_B(a_{ij}) + \pi_B(b_{ij})$ of the (i, j)th pixels

$$\frac{e^{\mu_A(a_{ij})+\pi_A(a_{ij})}}{e^{\mu_B(b_{ij})+\pi_B(b_{ij})}}$$

Corresponding to the fuzzy entropy, the divergence between images A and B due to $m_1(A)$ and $m_1(B)$ may be given as

$$D_1(A,B) = \sum_i\sum_j \left(1-(1-\mu_A(a_{ij})) \cdot e^{\mu_A(a_{ij})-\mu_B(b_{ij})} - \mu_A(a_{ij}) \cdot e^{\mu_B(b_{ij})-\mu_A(a_{ij})}\right) \quad (4.14)$$

Similarly, the divergence of B against A is

$$D_1(B,A) = \sum_i \sum_j \left(1-(1-\mu_B(b_{ij}))e^{\mu_B(b_{ij})-\mu_A(a_{ij})} - \mu_B(b_{ij})e^{\mu_A(a_{ij})-\mu_B(b_{ij})}\right) \quad (4.15)$$

So, the total divergence between pixels a_{ij} and b_{ij} of images A and B due to $m_1(A)$ and $m_1(B)$ is

$$Div-m_1(A,B) = D_1(A,B)+D_1(B,A)$$

$$= \sum_i \sum_j \left(2-(1-\mu_A(a_{ij})+\mu_B(b_{ij}))e^{\mu_A(a_{ij})-\mu_B(b_{ij})}\right.$$

$$\left. -(1-\mu_B(b_{ij})+\mu_A(a_{ij}))e^{\mu_B(b_{ij})-\mu_A(a_{ij})}\right) \quad (4.16)$$

Likewise, the total divergence between pixels a_{ij} and b_{ij} of images A and B due to $m_2(A)$ and $m_2(B)$ is

$$Div-m_2(A,B)$$

$$= \sum_i \sum_j \left(2-\left[1-(\mu_A(a_{ij})+\pi_A(a_{ij}))+(\mu_B(b_{ij})+(\pi_B(b_{ij}))\right]\cdot e^{\mu_A(a_{ij})+\pi_A(a_{ij})-(\mu_B(b_{ij})+\pi_B(b_{ij}))}\right.$$

$$\left. -\left[1-(\pi_B(b_{ij})+\mu_B(b_{ij}))+(\pi_A(a_{ij})+\mu_A(a_{ij}))\right]\cdot e^{\pi_B(b_{ij})+\mu_B(b_{ij})-(\mu_A(a_{ij})+\pi_A(a_{ij}))}\right)$$

$$(4.17)$$

Thus, the overall *IFD* between images A and B by adding Equations 4.16 and 4.17 is defined as

$$IFD(A,B)$$

$$= Div-m_1(A,B)+Div-m_2(A,B)$$

$$= \sum_i \sum_j \left(2-[1-\mu_A(a_{ij})+\mu_B(b_{ij})]e^{\mu_A(a_{ij})-\mu_B(b_{ij})} - [1-\mu_B(b_{ij})+\mu_A(a_{ij})]e^{\mu_B(b_{ij})-\mu_A(a_{ij})}\right.$$

$$+(2-[1-(\mu_A(a_{ij})-\mu_B(b_{ij}))+(\pi_B(b_{ij})-\pi_A(a_{ij}))]\cdot e^{\mu_A(a_{ij})-\mu_B(b_{ij})-(\pi_B(b_{ij})-\pi_A(a_{ij}))}$$

$$\left. -[1-(\pi_B(b_{ij})-\pi_A(a_{ij}))+(\mu_A(a_{ij})-\mu_B(b_{ij}))]\cdot e^{\pi_B(b_{ij})-\pi_A(a_{ij})-(\mu_A(a_{ij})-\mu_B(b_{ij}))}\right)$$

$$(4.18)$$

4.5 Intuitionistic Fuzzy Information Measure

If p and q are two probability distributions for two random variables, the cross-entropy measure of p and q is defined as [11]

$$I(p,q) = \sum p(x)\ln\frac{p(x)}{q(x)}$$

It measures the amount of discrimination of p and q. Lin [13] proposed a modified cross-entropy measure as

$$K(p,q) = \sum p(x)\ln\frac{p(x)}{(1/2)p(x)+(1/2)q(x)} \tag{4.19}$$

In an analogous manner, Vlachos and Sergiadis [20] defined information of discrimination in terms of the membership and non-membership degrees as

$$I(A,B) = \ln\frac{\mu_A(a_{ij})}{\mu_B(b_{ij})}$$

In order to define the cross entropy using *IFS*, information carried by both membership and non-membership degrees is calculated.
Information carried out as a result of the membership degree is

$$I'(A,B) = \sum \mu_A(x_i)\ln\frac{\mu_A(x_i)}{\mu_B(x_i)}$$

This is the expected information of discrimination of A against B. Likewise, the information due to the non-membership degree is

$$I''_{non}(A,B) = \sum v_A(x_i)\ln\frac{v_A(x_i)}{v_B(x_i)}$$

So, information of discrimination in favour of A against B [20] is

$$I'_{IFS}(A,B) = \sum_{i=1}^{n}\mu_A(x_i)\ln\frac{\mu_A(x_i)}{\mu_B(x_i)} + v_A(x_i)\ln\frac{v_A(x_i)}{v_B(x_i)}$$

If $\mu_B(x_i) = 0$ or $v_B(x_i) = 0$, the equation becomes undefined. So, in that case the equation is modified as

$$I_{IFS}(A,B) = \sum_{i=1}^{n} \mu_A(x_i) \ln \frac{\mu_A(x_i)}{(1/2)(\mu_A(x_i)+\mu_B(x_i))} + v_A(x_i) \ln \frac{v_A(x_i)}{(1/2)(v_A(x_i)+v_B(x_i))}$$

(4.20)

Symmetric information of the discrimination measure for *IFS* is

$$D_{IFS}(A,B) = I_{IFS}(A,B) + I_{IFS}(B,A)$$

4.6 Intuitionistic Fuzzy Entropy

Entropy is a measure of fuzziness in a fuzzy set. Zadeh [25] first introduced the idea of fuzzy entropy in 1969. Kaufmann [10] used the distance measure to define fuzzy entropy, whereas Yager [24] defined entropy as the distance from a fuzzy set and its complement. Similarly in the case of *IFS*, intuitionistic fuzzy entropy (*IFE*) gives the amount of vagueness or ambiguity in a set. Many authors defined *IFE* in a different manner. Two definitions of entropy of *IFS* were given by Burillo and Bustince [2] and Szmidt and Kacprzyk [17]. These two definitions have different frameworks. Burillo and Bustince defined entropy for the first time in terms of the degree of intuitionism of an *IFS*. Szmidt and Kacprzyk defined entropy in terms of the non-probabilistic type of entropy. In *IFS*, three parameters are taken into account with $\mu_A + v_A + \pi_A = 1$, $\mu_A \geq 0$, v_A, $\pi_A \leq 1$.

The properties of *IFE* by Burillo and Bustince [2] are
A real function *IFE* = *IFSs*(X) → [0, 1] is called *IFE* on *IFSs*(X) if

1. $IFE(A) = 0, \forall A \in FS(X)$.
2. $IFE(A) = \text{Cardinal}(X) = n$, iff $\mu_A(x_i) = v_A(x_i) = 0 \; \forall \; x_i$, that is, the entropy is maximum if the set is totally intuitionistic.
3. If the membership and non-membership of each element increase, their sum will increase; thereby, the fuzziness will increase and the entropy will decrease. Mathematically, it can be written as $IFE(A) \geq IFE(B)$ if $\mu_A(x_i) \leq \mu_B(x_i)$ and $v_A(x_i) \leq v_B(x_i)$.
4. $IFE(A) = IFE(A^c)$.

They defined entropy as

$$IFE(A) = \sum_{i=1}^{n} \pi_A(x_i)$$

4.6.1 Different Types of Entropies

1. Entropy is also defined by Chaira [4] that follows the previous norms and is given as

   For a probability distribution, $p = p_1, p_2, ..., p_n$, the exponential entropy is defined as $H = \sum_{i=1}^{n} p_i \cdot e^{(1-p_i)}$.

   For intuitionistic fuzzy cases, if $\mu_A(x_i)$, $\nu_A(x_i)$ and $\pi_A(x_i)$ are the membership, non-membership and hesitation degrees of the elements of the set $X = \{x_1, x_2, ..., x_n\}$, respectively, then *IFE*, which denotes the degree of intuitionism in fuzzy set, may be given as

   $$IFE(A) = \sum_{i=1}^{n} \pi_A(x_i) \cdot e^{[1-\pi_A(x_i)]} \qquad (4.21)$$

   where $\pi_A(x_i) = 1 - (\mu_A(x_i) + \nu_A(x_i))$.

   Szmidt and Kacprzyk [17] defined *IFE* in a different manner:
   A real function $IFE = IFSs(X) \rightarrow [0, 1]$ is called *IFE* on $IFSs(X)$ if

   a. $IFE(A) = 0$ if A is a crisp set, that is, $\mu_A(x_i) = 0$ or $\mu_A(x_i) = 1$ for all $x_i \in X$.

   b. $IFE(A) = 1$, if $\mu_A(x_i) = \nu_A(x_i)$ for all $x_i \in X$.

   c. $IFE(A) < IFE(B)$ if A is less fuzzy than B, that is, $\mu_A(x_i) \leq \mu_B(x_i)$ and $\nu_A(x_i) \geq \nu_B(x_i)$ for $\mu_B(x_i) \leq \nu_B(x_i)$ or $\mu_A(x_i) \geq \mu_B(x_i)$ and $\nu_A(x_i) \leq \nu_B(x_i)$ for $\mu_B(x_i) \geq \nu_B(x_i)$ for any $x_i \in X$.

   d. $IFE(A) = IFE(A^C)$.

2. Szmidt and Kacprzyk [17] also defined entropy as a ratio of the distances between an *IFS* and its nearest and farthest crisp sets in terms of cardinalities of an *IFS*. A brief note on the cardinalities of *IFS* is given later.

   The least cardinality or the sigma *count* is the min sigma *count* of A which is the arithmetic sum of the membership grades in a set and is also called here as min sigma *count*, which is given as

   $$\min \sum count(A) = \sum_{i=1}^{n} \mu_A(x_i)$$

   The biggest cardinality or max sigma *count* which is due to the hesitation degree, π_A, is given as

   $$\max \sum count(A) = \sum_{i=1}^{n} (\mu_A(x_i) + \pi_A(x_i)) \qquad (4.22)$$

Likewise, for A^c, the least cardinality or min sigma *count* is given as

$$\min \sum count(A^c) = \sum_{i=1}^{n} v_A(x_i)$$

The biggest cardinality or max sigma *count* which is due to the hesitation degree, π_A, is given as

$$\max \sum count(A^c) = \sum_{i=1}^{n} (v_A(x_i) + \pi_A(x_i))$$

Then cardinality of *IFS* is defined in an interval as

$$card\,A = \left[\min \sum count(A), \max \sum count(A)\right]$$

To explain the entropy as the ratio of the distances between an *IFS* and its nearest and farthest crisp sets, Figure 4.1 is used.

A and B correspond to a non-fuzzy/crisp set where point A corresponds to an element that fully belongs to 1, that is, $(\mu_A, v_A, \pi_A) = (1, 0, 0)$, and point B corresponds to an element that does not belong to the set $(\mu_B, v_B, \pi_B) = (0, 1, 0)$. Fuzzy set corresponds to line AB. When we move away along the line from point A to point B, the membership function decreases from 1 at point A to 0 at point B. At C, that is, at the midpoint, both the membership and non-membership

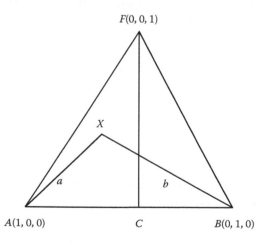

FIGURE 4.1
Graphical representation of IFS.

functions are 0.5, that is, the highest degree of fuzziness exists at this point. It is not known whether point C belongs to or does not belong to the set.

An *IFS* is represented by a triangle *ABF*. Points above line *AB* have a hesitation degree more than 0. Point F is not defined as the hesitation degree is equal to 1, and so it is difficult to say whether this point belongs/does not belong to the set. For *FC*, the membership degree $\mu(FC)$ and non-membership degree $\nu(FC)$ are equal, but the hesitation degree is greater than 0, and the condition still follows: $\mu_{FC}(x_i) + \nu_{FC}(x_i) + \pi_{FC}(x_i) = 1$. So, for every point x_i in *FC*, $dist(A, x_i) = dist(B, x_i)$ follows.

The entropy of *IFS* is based on ratio-based measure:

$$E(X) = \frac{a}{b} \tag{4.23}$$

where

$a = dist(X, X_{near})$ means the distance from X to the nearest point X_{near} from A and B

$b = dist(X, X_{far})$ is the distance from X to the farthest point X_{far} from A and B

This equation is the entropy for one point. From this entropy, it can be said that either X fully belongs to point A or does not belong to point B.

For 'n' points, the entropy in Equation 4.23 is redefined as

$$IFE = \frac{1}{n}\sum_{i=1}^{n} E(X_i) \tag{4.24}$$

3. The generalized entropy measure of set A of 'n' elements in terms of max sigma *count* is written as [17]

$$IFE(A) = \frac{1}{n}\sum_{i=1}^{n} \frac{\max count\left(A_i \cap A_i^c\right)}{\max count\left(A_i \cup A_i^c\right)} \tag{4.25}$$

where

$$A_i \cap A_i^c = \left\langle \min\left(\mu_{Ai}, \mu_{Ai}^c\right), \max\left(\nu_{Ai}, \nu_{Ai}^c\right)\right\rangle$$

$$A_i \cup A_i^c = \left\langle \max\left(\mu_{Ai}, \mu_{Ai}^c\right), \min\left(\nu_{Ai}, \nu_{Ai}^c\right)\right\rangle$$

4. Szmidt and Kacprzyk [17] suggested a similar type of entropy which is given as

$$IFE(A) = \frac{1}{n}\sum_{i=1}^{n}\frac{\min\{(\mu_A(x_i),\nu_A(x_i)\}+\pi_A(x_i)}{\max\{(\mu_A(x_i),\nu_A(x_i)\}+\pi_A(x_i)} \qquad (4.26)$$

5. An entropy suggested by Huang and Liu [7] on a vague set but extended to *IFS* is defined as

$$IFE(A) = \frac{1}{n}\sum_{i=1}^{n}\frac{1-|\mu_A(x_i)-\nu_A(x_i)|+\pi_A(x_i)}{1+|\mu_A(x_i)-\nu_A(x_i)|+\pi_A(x_i)} \qquad (4.27)$$

6. Vlachos and Sergiadis [19] introduced an entropy which is defined as

$$IFE(A) = \frac{1}{n}\sum_{i=1}^{n}\frac{2\mu_A(x_i)\cdot\nu_A(x_i)+\pi_A^2(x_i)}{\mu_A^2(x_i)+\nu_A^2(x_i)+\pi_A^2(x_i)}$$

Two different kinds of measures using *IFS* are given by Ye [23]:

$$E_{IFS}(A)$$

$$=\frac{1}{n}\sum_{i=1}^{n}\left[\left\{\sin\frac{\pi\times[1+\mu_A(x_i)-\nu_A(x_i)]}{4}+\sin\frac{\pi\times[1-\mu_A(x_i)+\nu_A(x_i)]}{4}-1\right\}\times\frac{1}{\sqrt{2}-1}\right]$$

$$E_{IFS}(A)$$

$$=\frac{1}{n}\sum_{i=1}^{n}\left[\left\{\cos\frac{\pi\times[1+\mu_A(x_i)-\nu_A(x_i)]}{4}+\cos\frac{\pi\times[1-\mu_A(x_i)+\nu_A(x_i)]}{4}-1\right\}\times\frac{1}{\sqrt{2}-1}\right]$$

4.7 Entropy of Interval-Valued Intuitionistic Fuzzy Set

In *IFS* A, $\mu_A(x)$ and $\nu_A(x)$ denote the membership and non-membership functions, respectively. For convenience, $\mu_A(x)$ may be represented as $\mu_A(x)=[\mu_A^-(x),\mu_A^+(x)]$ and $\nu_A(x)$ may be represented as $\nu_A(x)=[\nu_A^-(x),\nu_A^+(x)]$. So, interval-valued intuitionistic fuzzy set (*IVIFS*) is represented as $A=\{x,[\mu_A^-(x),\mu_A^+(x)],[\nu_A^-(x),\nu_A^+(x)]|,x\in X\}$.

The interval $[\pi_A^-(x), \pi_A^+(x)] = [1 - \mu_A^-(x) - v_A^-(x), 1 - \mu_A^+(x) - v_A^+(x)]$ is the hesitation degree.

A real function E: $IVIFS(X) \rightarrow (0, 1)$ is called the entropy of $IVIFS$ set if it satisfies the following properties [22]:

1. $E(A) = 0$ if A is a crisp set.
2. $E(A) = 1$ if $[\mu_A^-(x), \mu_A^+(x)] = [v_A^-(x), v_A^+(x)]$ for all $x_i \in X$.
3. $E(A) \leq E(B)$ if $A \subseteq B$ when $\mu_B^-(x) \leq v_B^-(x)$ and $\mu_B^+(x) \leq v_B^+(x)$ or $B \subseteq A$ if $\mu_B^-(x) \geq v_B^-(x)$ and $\mu_B^+(x) \geq v_B^+(x)$ for any $x_i \in X$.

As in the entropy of *IFS*, the entropy of *IVIFS* is given as [22]

$$E_{IVIFS}(A) = \frac{1}{n} \sum_{i=1}^{n} \frac{\min\{\mu_A^-(x_i), v_A^-(x_i)\} + \min\{\mu_A^+(x_i), v_A^+(x_i)\} + \pi_A^-(x_i) + \pi_A^+(x_i)}{\max\{\mu_A^-(x_i), v_A^-(x_i)\} + \max\{\mu_A^+(x_i), v_A^+(x_i)\} + \pi_A^-(x_i) + \pi_A^+(x_i)}$$

(4.28)

Another type of entropy by Zhang and Jiang [26] is given as

$$E_{IVIFS}(A) = \frac{\sum_{i=1}^{n} \left(\mu_A^-(x_i) \wedge v_A^-(x_i) + \mu_A^+(x_i) \wedge v_A^+(x_i) \right)}{\sum_{i=1}^{n} \left(\mu_A^-(x_i) \vee v_A^-(x_i) + \mu_A^+(x_i) \vee v_A^+(x_i) \right)}$$

(4.29)

$$E_{IVIFS}(A) = 1 - \frac{1}{n} \sum_{i=1}^{n} \left(\left| \mu_A^-(x_i) - v_B^-(x_i) \right| \vee \left| \mu_A^+(x_i) - v_B^+(x_i) \right| \right)$$

4.8 Similarity Measure and Distance Measures of *IVIFS*

For any two *IVIFS*s A and B,

$$A^{IVIFS} = \left\{ x, \left[\mu_A^-(x), \mu_A^+(x) \right], \left[v_A^-(x), v_A^+(x) \right] \middle| x \in X \right\}$$

$$B^{IVIFS} = \left\{ x, \left[\mu_B^-(x), \mu_B^+(x) \right], \left[v_B^-(x), v_B^+(x) \right] \middle| x \in X \right\}$$

the similarity measure is defined as [22] from the entropy in Equation 4.27, which is written as

$$S_{IVIFS}(A,B)$$

$$= \frac{1}{n} \sum_{i=1}^{n} \frac{2 - \min\left\{ \left|\mu_{\bar{A}}(x_i) - \mu_{\bar{B}}(x_i)\right|, \left|v_{\bar{A}}(x_i) - v_{\bar{B}}(x_i)\right| \right\} - \min\left\{ \left|\mu_{A}^{+}(x_i) - \mu_{B}^{+}(x_i)\right|, \left|v_{A}^{+}(x_i) - v_{B}^{+}(x_i)\right| \right\}}{2 + \max\left\{ \left|\mu_{\bar{A}}(x_i) - \mu_{\bar{B}}(x_i)\right|, \left|v_{\bar{A}}(x_i) - v_{\bar{B}}(x_i)\right| \right\} + \max\left\{ \left|\mu_{A}^{+}(x_i) - \mu_{B}^{+}(x_i)\right|, \left|v_{A}^{+}(x_i) - v_{B}^{+}(x_i)\right| \right\}}$$

$$\forall A, B \in IVIFS(A)$$

The normalized Euclidean similarity measure induced by the Hausdorff metric [15] is given as

$$S_{IVIFS}(A,B) = 1 - \left\{ \frac{1}{2n} \sum_{i=1}^{n} \left[\left(\left|\mu_{\bar{A}}(x_i) - \mu_{\bar{B}}(x_i)\right| \vee \left|\mu_{A}^{+}(x_i) - \mu_{B}^{+}(x_i)\right| \right)^2 \right.\right.$$

$$\left.\left. + \left(\left|v_{\bar{A}}(x_i) - v_{\bar{B}}(x_i)\right| \vee \left|v_{A}^{+}(x_i) - v_{B}^{+}(x_i)\right| \right)^2 \right] \right\}^{1/2}$$

The distance measure using Hamming and Euclidean distance [15] is given as

$$d_{IVIFS}(a,b)$$

$$= \frac{1}{4} \sum_{i=1}^{n} \left[\left|\mu_{\bar{A}}(x_i) - \mu_{\bar{B}}(x_i)\right| + \left|\mu_{A}^{+}(x_i) - \mu_{B}^{+}(x_i)\right| + \left|v_{\bar{A}}(x_i) - v_{\bar{B}}(x_i)\right| + \left|v_{A}^{+}(x_i) - v_{B}^{+}(x_i)\right| \right]$$

$$d_{IVIFS}(a,b) = \frac{1}{4} \sum_{i=1}^{n} \left[\left(\mu_{\bar{A}}(x_i) - \mu_{\bar{B}}(x_i)\right)^2 + \left(\mu_{A}^{+}(x_i) - \mu_{B}^{+}(x_i)\right)^2 \right.$$

$$\left. + \left(v_{\bar{A}}(x_i) - v_{\bar{B}}(x_i)\right)^2 + \left(v_{A}^{+}(x_i) - v_{B}^{+}(x_i)\right)^2 \right]^{1/2}$$

4.9 Summary

As is well known that entropy and similarity measures are two important issues in fuzzy set theory, they are widely used in image processing, pattern recognition, cluster analysis and so on. But in real-life situations, there are many uncertainties where human negation does not satisfy logical negation, and so in this chapter, the definition, properties and different types of distance, similarity, entropy and information measures using *IFS* and *IVIFS* are discussed. These are very useful especially in medical image processing where the presence of uncertainties is high.

References

1. Bhandari, D. and Pal, N.R., Some new information measures for fuzzy sets, *Information Sciences*, 67, 209–228, 1993.
2. Burillo, P. and Bustince, H., Entropy on intuitionistic fuzzy sets and on interval-valued fuzzy sets, *Fuzzy Sets and Systems*, 78, 305–316, 1996.
3. Chaira, T. and Ray, A.K., A new measure using intuitionistic fuzzy set theory and its application to edge detection, *Applied Soft Computing*, 2, 919–927, 2008.
4. Chaira, T., A novel intuitionistic fuzzy c means clustering algorithm and its application to medical images, *Applied Soft Computing*, 11(2), 1711–1717, 2011.
5. Dengfeng, L. and Chuntian, C., New similarity measure of intuitionistic fuzzy sets and application to pattern recognitions, *Pattern Recognition Letters*, 23, 221–225, 2002.
6. Grzegorzewski, P., Distances between intuitionistic fuzzy sets and/or interval-valued fuzzy sets based on Hausdorff metric, *Fuzzy Sets and Systems*, 128(2), 319–328, 2004.
7. Huang, G.S. and Liu, Y.S., The fuzzy entropy of vague sets based on non-fuzzy sets, *Computer Applications and Software*, 22(6), 16–17, 2005.
8. Hung, W.-L. and Yang, M.-S., Similarity measures of intuitionistic fuzzy sets based on Hausdorff distance, *Pattern Recognition Letters*, 25, 1603–1611, 2004.
9. Hung, W.L. and Yang, M.S., On similarity measures between intuitionistic fuzzy sets, *International Journal of Intelligent Systems*, 23, 364–383, 2008.
10. Kaufmann, A., *Introduction to the Theory of Fuzzy Subsets: Fundamental Theoretical Elements*, Vol. 1, Academic Press, New York, 1980.
11. Kullback, S., *Information Theory and Statistics*, Dover Publications, New York, 1968.
12. Liang, D. and Shi, C., New similarity measures of intuitionistic fuzzy sets and application to pattern recognition, *Pattern Recognition Letters*, 23, 221–225, 2002.
13. Lin, J., Divergence measures based on the Shannon entropy, *IEEE Transactions on Information Theory*, 37, 145–151, 1991.
14. Nadler Jr., S.B., *Hyperspaces of Sets*, Marcel Dekker, New York, 1978.
15. Park, J.H. et al., Distances between interval-valued intuitionistic fuzzy sets, *Journal of Physics: Conference Series*, 96, 1–8, 2008.
16. Song, Y.-T. and Zhou, X.-G., New properties and measures of distance measure between intuitionistic fuzzy sets, in *Proc. of IEEE Conference on Fuzzy Systems and Knowledge Discovery*, Tianjin, China, pp. 40–343, 2009.
17. Szmidt, E. and Kacprzyk, J., Entropy for intuitionistic fuzzy set, *Fuzzy Sets and Systems*, 118, 467–477, 2001.
18. Szmidt, E. and Kacprzyk, J., Distance between intuitionistic fuzzy set, *Fuzzy Sets and Systems*, 118, 505–518, 2000.

19. Vlachos, I.K. and Sergiadis, G.D., Inner product based entropy in the intuitionistic fuzzy setting, *International Journal of Uncertainty Fuzziness Knowledge Based Systems*, 14, 351–366, 2006.
20. Vlachos, I.K. and Sergiadis, G.D., Intuitionistic fuzzy information – Applications to pattern recognition, *Pattern Recognition Letters*, 28, 197–206, 2007.
21. Wang, W. and Xin, X., Distance measure between intuitionistic fuzzy sets, *Pattern Recognition Letters*, 26, 2063–2069, 2005.
22. Wei, C.-P., Wang, P., and Zhang, Y.-Z., Entropy, similarity measure of interval-valued intuitionistic fuzzy sets and their applications, *Information Sciences*, 181, 4273–4286, 2011.
23. Ye, J., Two effective measures of intuitionistic fuzzy entropy, *Computing*, 87, 55–62, 2010.
24. Yager, R.R., On the measure of fuzziness and negation. I. Membership in unit interval, *International Journal of General Systems*, 5, 221–229, 1979.
25. Zadeh, L.A., Fuzzy sets, *Information and Control*, 8, 338–353, 1965.
26. Zhang, Q. and Jiang, S., Relationships between entropy and similarity measure of interval-valued intuitionistic fuzzy sets, *International Journal of Intelligent Systems*, 25, 1121–1140, 2010.

5

Image Enhancement

5.1 Introduction

Image enhancement is required before processing any image. It plays a fundamental role in image processing where human experts make important decisions based on image information. It is used to restore an image that has deteriorated or to enhance certain features of an image. The reason for image enhancement is to transform an image to another form that is more suitable for further processing. For visual analysis of a medical image, physicians should have a good knowledge on the images of the patient for better diagnosis. Medical images are poorly illuminated, and so many structures are not clearly visible. Many regions/boundaries are vague/fuzzy in nature. So, if the quality of the image is improved, processing may become easier. For this reason, medical image enhancement is extremely important. In an enhanced image, it becomes easier for specialists or doctors to spot the anomalies in x-rays, CT scan or mammogram or in pathological images or in any other images. Image enhancement includes contrast enhancement and edge enhancement.

The purpose of contrast enhancement is to increase the overall visual contrast of the image, which the human eye can visualize clearly, to be better suited for further analysis. It is useful when the intensity of important regions of images such as tissues, blood vessels or fine structures in medical images is very low, and it becomes very difficult to make out the structures with the human eye. Contrast enhancement highlights the areas of low intensity, thus improving the readability. Edge enhancement highlights the edges of the abnormal lesions or any structures in the images, especially the images where the edges are not clearly visible.

There are many crisp methods on image enhancement, and the most common method is histogram equalization. Other methods include grey-level transformation or grey-level modification. But medical images contain uncertainties, so crisp enhancement may not improve the image properly. To deal with such kind of images, different types of mathematical tools such as fuzzy set theory and some advanced fuzzy set theories are suggested by many authors who deal with uncertainties in a different manner. There are many fuzzy methods that manage vague data and that perform better. In some cases, that is, in real-time images, fuzzy enhancement does not provide satisfactory results. This may be due to the fact that fuzzy methods consider only one uncertainty which is in the form of a membership function. So, advanced fuzzy set theories, such as intuitionistic fuzzy (*IF*) set and Type II fuzzy set, that consider more uncertainties are used in image enhancement to obtain better results.

Fuzzy methods for image enhancement are already described in my first book *Fuzzy Image Processing and Application in MATLAB*, so fuzzy enhancement is not discussed in detail in this chapter. A brief overview on fuzzy enhancement of medical images along with different fuzzy methods is presented in the next section.

5.2 Fuzzy Image Contrast Enhancement

Contrast is a property that is based on human perception. An approximate definition of contrast is

$$C = \frac{(A-B)}{(A+B)}$$

where A and B are the mean grey levels of the two regions where the contrast is calculated.

Contrast enhancement is applied to the images where the contrast between the object and the background is very low, that is, when the objects are not properly differentiable from the background. In this case, contrast enhancement should be such that darker regions should appear darker and lighter regions should appear lighter, but no contrast enhancement is required when the contrast of the image is better. Fuzzy image contrast enhancement is based on grey-level mapping from a crisp (grey) plane into a fuzzy plane using a certain membership transformation. Based on a user-defined threshold, T, the contrast is stretched in such a way that the grey levels below the threshold T are reduced and the grey levels above the threshold T are increased in a non-linear fashion. This stretching operation induces saturation at both ends (grey levels) [1]. The idea may be extended for multiple thresholds where different regions are stretched in a different fashion depending on the quality of an image. Fuzzy contrast enhancement initially assigns membership values $\mu(x)$ that may be triangular, Gaussian, gamma, etc., to know the degree of brightness or

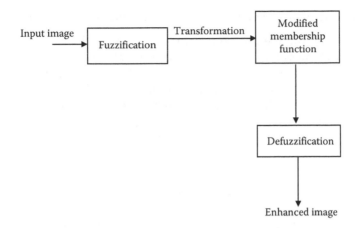

FIGURE 5.1
Fuzzy image enhancement.

darkness of the pixels in an image. Then a transformation function is applied on the membership values to generate new membership values of the pixels in the image. Finally, an inverse transformation is applied on the new membership values for transforming back the membership values to a spatial domain. The principle of fuzzy contrast enhancement is illustrated in Figure 5.1.

Algorithmically, it can be expressed as

$$\mu'(x) = \psi(\mu(x))$$

$$x' = f^{-1}(\mu'(x))$$

where
$\mu(x)$ is the membership function
$\psi(\mu(x))$ is the transformation of $\mu(x)$ denoted as $\mu'(x)$

In recent years, many researchers [12,13] have applied various fuzzy methods for contrast enhancement. Before discussing the use of advanced fuzzy set theoretic techniques in image enhancement, a few fuzzy methods are discussed briefly in the next section.

5.3 Fuzzy Methods in Contrast Enhancement

In this section, different fuzzy methods for contrast enhancement are discussed briefly.

5.3.1 Contrast Enhancement Using the Intensification Operator

In this method, the membership values are modified using an intensifier. Initially, the membership function is selected that finds the membership values of the pixels of an image. Then the transformation of the membership values which are above 0.5 (default value) to much higher values and the membership values which are lower than 0.5 to much lower values is carried out in a non-linear fashion to obtain a good contrast in an image. The contrast intensifier (INT) operation is written as

$$\mu'_{mn} = \frac{2 \cdot [\mu_{mn}]^2}{1 - 2 \cdot [1 - \mu_{mn}]^2} \quad \begin{array}{l} 0 \leq \mu_{mn} \leq 0.5 \\ 0.5 \leq \mu_{mn} \leq 1 \end{array}$$

Once the membership values are modified, modified grey values are then transformed to spatial domain using the inverse function. There is also an operator called 'NINT' which uses Gaussian membership function in membership function generation [12].

5.3.2 Contrast Improvement Using Fuzzy Histogram Hyperbolization

The concept of fuzzy histogram hyperbolization was discussed by Tizhoosh and Fochem [16]. This method modifies the membership values of the grey levels into the logarithmic function due to non-linear human brightness perception. Initially, a membership function is selected that finds the membership values of the pixels of an image. A fuzzifier beta, β, which is a linguistic hedge, is set to modify the membership function. Hedges [10,15] may be very bright, medium bright, etc., and the selection is made on the basis of the user's needs. The value of beta may be in the range $\beta \in [0.5, 2]$. Depending on the value of β, the operation may be dilation or concentration. If the image is a low-intensity image, then the fuzzifier β after operating on the membership values will produce slightly bright or quiet bright images.

5.3.3 Contrast Enhancement Using IF-THEN Rules

Image quality can be improved by using human knowledge, which is highly subjective in nature. Different observers judge the image differently. The fuzzy rule–based approach is such a method that incorporates human intuitions which are non-linear in nature, and these cannot be easily characterized by traditional modelling. As it is hard to define a precise or crisp condition under which enhancement is applied, the fuzzy set theoretic approach is a good approach to this solution. The rule-based approach incorporates fuzzy rules into the conventional methods. A set of conditions on the pixel that are related to the pixel grey level and also the pixel neighbourhood (if requires) are defined, and these conditions will form the antecedent part of the IF-THEN rules. Fuzzy rule–based systems make soft decisions on each condition, aggregates the decision made and finally makes a decision based on the aggregation.

5.3.4 Contrast Improvement Using the Fuzzy Expected Value

The use of the fuzzy expected value (FEV) in image enhancement was proposed by Schneider and Craig [14] where the image quality is improved in terms of the distance between the grey levels and the FEV. The FEV by Friedman et al. [9] replaces the mean and median value with a more representative value or 'typical value' when treating with fuzzy sets. This value would indicate a typical grade of membership of a fuzzy set.

Example 5.1

Two examples are shown in Figures 5.2 and 5.3 to show the contrast enhancement of blood vessels using four fuzzy methods for image enhancement on low-contrast blood vessel images.

As it is difficult to say which method will perform the best, researches are carried out to improve the quality of contrast of the resultant images. Fuzzy methods do provide better results, but in some cases, they fail to provide better results. Intuitionistic fuzzy (IF) set and Type II fuzzy set theoretic techniques are used bearing in mind that as these sets consider more uncertainties, better results may be obtained. So, when fuzzy methods in some cases fail to provide better results, advanced fuzzy techniques may be used.

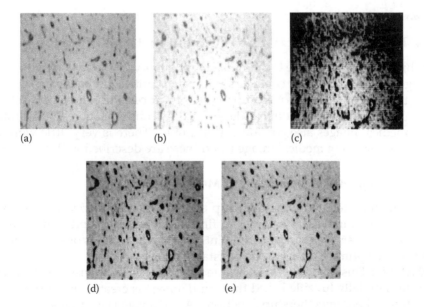

(a)　　　　　(b)　　　　　(c)

(d)　　　　　(e)

FIGURE 5.2
(a) Blood vessel (BV1) image, (b) enhancement using the FEV, (c) enhancement using the INT operator, (d) enhancement using the NINT operator and (e) enhancement using histogram hyperbolization.

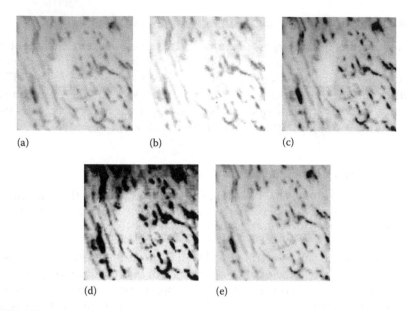

(a)   (b)   (c)

(d)   (e)

FIGURE 5.3
(a) Blood vessel (BV2) image, (b) enhancement using the FEV, (c) enhancement using the INT operator, (d) enhancement using the NINT operator and (e) enhancement using histogram hyperbolization.

5.4 Intuitionistic Fuzzy Enhancement Methods

In this section to improve the enhancement quality, *IF* methods are used for image enhancement as it considers two uncertainties: the membership and non-membership degrees as compared to one uncertainty in a fuzzy set. So, it is expected to obtain better results as *IF* [2] set considers more uncertainties and medical images also contain uncertainties. There is very little work on *IF* enhancement of medical images, and these are described with examples.

5.4.1 Entropy-Based Enhancement Methods

In *IF* enhancement, both membership and non-membership values of an *IF* image are required to determine. To find the degrees, an optimum value of the constant parameter is required. Entropy-based methods used *IF* entropy to find the optimum value of the constant term.

Method I: This method was suggested by Vlachos and Sergiadis [17]. The image is initially fuzzified, and then an *IF* image is created using the membership and non-membership functions. An *IF* image is written as

$$A_{IFS} = \{x, \mu_A(g), \nu_A(g)\}, \quad g \in \{0,1,2,\dots,L-1\}$$

where g is the pixel value.

An image (say A) of size $M \times N$ is initially fuzzified using the following formula:

$$\mu_A(g) = \frac{g - g_{min}}{g_{max} - g_{min}} \tag{5.1}$$

where
  g is the grey level that ranges from 0 to L −a1
  g_{min} and g_{max} are the minimum and maximum values of the grey levels of the image, respectively

Based on the fuzzy set, the membership degree of the *IF* image is calculated as

$$\mu_{IFS}(g; \lambda) = 1 - (1 - \mu_A(g))^{\lambda - 1}$$

Using standard fuzzy negation, $\varphi(x) = (1 - x)^\lambda$, $\lambda \geq 1$, the non-membership function is given as

$$\nu_{IFS}(g; \lambda) = \varphi(\mu_{IFS}(g; \lambda)) \tag{5.2}$$

or

$$\nu_{IFS}(g; \lambda) = (1 - \mu_{IFS}(g; \lambda))^\lambda, \quad \lambda \geq 1$$
$$= (1 - \mu_A(g; \lambda))^{\lambda(\lambda - 1)} \tag{5.3}$$

The hesitation degree is

$$\pi_{IFS}(g; \lambda) = 1 - \mu_{IFS}(g; \lambda) - \nu_{IFS}(g; \lambda)$$

As λ is not fixed for all the images, the optimum value of λ is obtained using *IF* entropy. The optimum values are calculated using different entropies. There are many types of entropies suggested by different authors. These are

1. Entropy by Burillo and Bustince [4]

$$E_1(A_{IFS}) = \frac{1}{M \times N} \sum_{j=0}^{M-1} \sum_{i=0}^{N-1} (1 - \mu_A(g_{ij}) - \nu_A(g_{ij})) e^{1 - (\mu_A(g_{ij}) + \nu_A(g_{ij}))} \tag{5.4}$$

2. Entropy by Vlachos and Sergiadis [17]

$$E_2(A_{IFS}) = \frac{1}{M \times N} \sum_{j=0}^{N-1} \sum_{i=0}^{M-1} \frac{2\mu_A(g_{ij})\nu_A(g_{ij}) + \pi_A^2(g_{ij})}{\pi_A^2(g_{ij}) + \mu_A^2(g_{ij}) + \nu_A^2(g_{ij})} \tag{5.5}$$

3. Entropy by Burillo and Bustince

$$E_3(A_{IFS}) = \frac{1}{M \times N} \sum_{j=0}^{N-1} \sum_{i=0}^{M-1} \pi_A(g_{ij}) e^{(1-\pi_A(g_{ij}))}$$

Intuitionistic fuzzy entropy (*IFE*) is calculated from any of the entropies for all the λ values. The optimum value of λ that corresponds to the maximum value of the entropy values is written as

$$\lambda_{opt} = \max(IFE(A_{IFS}; \lambda))$$

So, in the *IF* domain, the image is represented as

$$A_{IFS\_opt} = \left\{ g, \mu_A(g; \lambda_{opt}), \nu_A(g; \lambda_{opt}) \middle| g \in 0, 1, \dots, L-1 \right\}$$

Atanassov's operator is applied to A_{IFS_opt} to deconstruct an *IF* image to a fuzzy image. With different values of α, different images are obtained in the fuzzy domain. Atanassov's operator is written as [3]

$$D_\alpha\left(A_{IFS\_opt}\right) = \left\{ x, \mu_A(x) + \alpha\pi_A(x), \nu_A(x) + (1-\alpha)\pi_A(x) \middle| x \in X \right\}, \quad \alpha \in [0,1]$$

The maximum index of fuzziness intuitionistic defuzzification [20] is used to select the optimum value of α. In computing the maximum index of defuzzification, the linear index of fuzziness is required.

The linear index of fuzziness of fuzzy set *A* is

$$\gamma_l(x) = \frac{1}{2|X|} \sum_{i=1}^{n} \min(\mu_A(x_i), 1 - \mu_A(x_i))$$

where $|X|$ is the cardinality of X, $n = |X|$. Substituting min *t*-norm with the product operator, the modified index of fuzziness is written as

$$\gamma_i(x) = \frac{1}{2|X|} \sum_{i=1}^{n} \mu_A(x_i)(1 - \mu_A(x_i)) \tag{5.6}$$

To find α_{opt}, the maximization index of fuzziness is required:

$$\alpha_{opt} = \max_{\alpha \in [0,1]} \left\{ \gamma(D_\alpha(A_{IFS\_opt})) \right\} \tag{5.7}$$

where

$$\gamma D_\alpha(A_{IFS\_opt}) = \frac{1}{4MN}\sum_{g=0}^{L-1}h_A(g)\Big(\mu_{D\alpha(A_{IFS\_opt})}(g)\cdot\big(1-\mu_{D\alpha(A_{IFS\_opt})}(g)\big)\Big)$$

$$= \frac{1}{4MN}\sum_{g=0}^{L-1}h_A(g)\big(\mu_A(g;\lambda_{opt})+\alpha\cdot\pi_A(g;\lambda_{opt})\big)\big(1-\mu_A(g;\lambda_{opt})-\alpha\cdot\pi_A(g;\lambda_{opt})\big)$$

Now differentiating $\gamma D_\alpha(A_{IFS\_opt})$ and equating it to 0,

$$\frac{d\gamma}{d\alpha} = \sum_{g=0}^{L-1}h_A(g)\big(\mu_A(g;\lambda_{opt})+\alpha\cdot\pi_A(g;\lambda_{opt})\big)\times\big(-\pi_A(g;\lambda_{opt})\big)$$

$$+ \sum_{g=0}^{L-1}h_A(g)\cdot\big(1-\mu_A(g;\lambda_{opt})-\alpha\cdot\pi_A(g;\lambda_{opt})\big)\times\pi_A(g;\lambda_{opt})=0$$

$$\Rightarrow \sum_{g=0}^{L-1}h_A(g)\big(\mu_A(g;\lambda_{opt})+\alpha\cdot\pi_A(g;\lambda_{opt})\big)\cdot\pi_A(g;\lambda_{opt})$$

$$= \sum_{g=0}^{L-1}h_A(g)\big(1-\mu_A(g;\lambda_{opt})-\alpha\cdot\pi_A(g;\lambda_{opt})\big)\cdot\pi_A(g;\lambda_{opt})$$

$$\Rightarrow \sum_{g=0}^{L-1}h_A(g)\cdot\pi_A(g;\lambda_{opt})-2\sum_{g=0}^{L-1}h_A(g)\cdot\mu_A(g;\lambda_{opt})\cdot\pi_A(g;\lambda_{opt})$$

$$= 2\alpha\sum_{g=0}^{L-1}h_A(g)\cdot\pi_A^2(g;\lambda_{opt})$$

or

$$\alpha'_{opt} = \frac{\sum_{g=0}^{L-1}h_A(g)\pi_A(g;\lambda_{opt})(1-2\mu_A(g;\lambda_{opt}))}{2\sum_{g=0}^{L-1}h_A(g)\pi_A^2(g;\lambda_{opt})}$$

As α'_{opt} may not lie in [0, 1], α_{opt} is used to keep the parameter in [0, 1]:

$$\alpha_{opt} = \alpha'_{opt} \quad \begin{matrix} 0 & \text{if } \alpha'_{opt} < 0 \\ & \text{if } 0 \le \alpha'_{opt} \le 1 \\ 1 & \text{if } \alpha'_{opt} > 1 \end{matrix}$$

where h_A being the histogram of the fuzzified image $\bar{A}$.

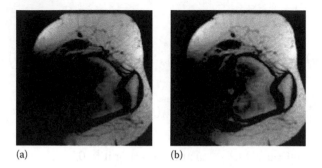

(a)   (b)

FIGURE 5.4
(a) Blood vessel image and (b) enhancement using the *IF* method (method I).

Finally, the image in the grey-level domain is written as

$$g' = (L-1)\mu_{D_{\alpha\_opt}(A_{opt})}(g)$$

and

$$\mu_{D_\alpha(A_{IFS\_opt})}(g) = \alpha_{opt} + (1-\alpha_{opt})\mu_A(g;\lambda_{opt}) - \alpha_{opt}\nu_A(g;\lambda_{opt}) \qquad (5.8)$$

g and g' are the initial and final intensity levels of the image, respectively.

Example 5.2

An example of a medical image using the *IF* method by Vlachos is shown in Figure 5.4 to illustrate the efficacy of the method. Entropy by Vlachos and Sergiadis is used in the enhancement method.

5.4.2 Two-Dimensional Entropy–Based Intuitionistic Fuzzy Enhancement (Method II)

The 2D entropy-based *IF* enhancement is suggested by Vlachos and Sergiadis [19] where they used two parameters instead of one. The image is fuzzified with the membership function $\mu_A(g)$ using Equation 5.1.

Using Yager's fuzzy complement, the membership function of the *IF* image (*A*) is given as

$$\mu_A^{IFS}(g;\omega) = \left(1-\mu_A^\omega(g)\right)^{1/\omega}, \quad \omega > 0 \qquad (5.9)$$

and the non-membership function is generated as

$$\nu_A^{IFS}(g;\omega,\lambda) = \varphi\left(\mu_A^{IFS}(g;\omega,\lambda)\right) = \left(1-\mu_A^\omega(g)\right)^{\lambda/\omega} \qquad (5.10)$$

This is obtained using the *IF* generator:

$$\varphi(x) = (1-x)^\lambda, \quad \lambda \geq 1, \quad x \in [0,1]$$

By varying λ and ω, different representations of images are obtained in the *IF* domain. To obtain the optimum values of ω and λ, the *IF* entropy is used:

$$E_A(A;\omega,\lambda) = \frac{1}{MN} \sum_{g=0}^{L-1} h_A(g) \frac{1-\max\left\{1-\left(1-\mu_A^\omega(g)\right)^{1/\omega}, \left(1-\mu_A^\omega(g)\right)^{\lambda/\omega}\right\}}{1-\min\left\{1-\left(1-\mu_A^\omega(g)\right)^{1/\omega}, \left(1-\mu_A^\omega(g)\right)^{\lambda/\omega}\right\}} \quad (5.11)$$

Entropy is a function of ω and λ. For a constant ω, $E_A(A; \omega, \lambda)$ attains a maximum for a specific value of λ denoted as $\lambda_{opt}(\omega)$. The optimum value is obtained by maximizing the fuzzy entropy. An *IF* image is obtained as

$$A\left(\lambda_{opt}(\omega),\omega\right) = \left\{x, \mu_A(g;\omega), v_A\left(g;\lambda_{opt},\omega\right) \middle| g \in 0,1,2,\ldots,L-1\right\}$$

$h_A(g)$ is the histogram of the fuzzified image.

5.4.3 Entropy-Based Enhancement Method by Chaira (Method III)

The entropy-based enhancement method is suggested by Chaira [6] where the image is considered fuzzy and so grey levels are imprecise. Due to quantization noise, a grey level g in a digital image may be $(g + 1)$ or $(g - 1)$. Taking this into account, for each grey level, g, the grey values that are $(g + 1)$ or $(g - 1)$ are replaced by grey level g. This is computed for all the grey levels.
   The image is initially fuzzified $\mu_A(g)$ using Equation 5.1.
   From Sugeno's fuzzy complement, the *IF* membership function is given as

$$\mu_A^{IFS}(g) = 1 - \frac{1-\mu_A(g)}{1+\lambda \cdot \mu_A(g)} = \frac{(1+\lambda) \cdot \mu_A(g)}{1+\lambda \cdot \mu_A(g)} \quad (5.12)$$

Using Sugeno's fuzzy negation,

$$\varphi(x) = \frac{1-x}{1+\lambda \cdot x} \quad (5.13)$$

While computing the non-membership function of an *IF* image, λ in Equation 5.13 is changed (increased) to $\lambda + 1$. With the change in λ, the non-membership degree, $v_A^{IFS}(g)$, will change (decrease) but will still follow the condition

$v_A^{IFS}(g) \le 1 - \mu_A^{IFS}(g)$. This is done to obtain a better contrast-enhanced image. The modified non-membership function is given as

$$v_A^{IFS}(g) = \phi(\mu_A^{IFS}(g)) = \frac{1 - \mu_A^{IFS}(g)}{1 + (\lambda + 1)\mu_A^{IFS}(g)} = \frac{1 - \dfrac{(1+\lambda)\mu_A(g)}{1 + \lambda \cdot \mu_A(g)}}{1 + \dfrac{(\lambda + 1)(1 + \lambda)\mu_A(g)}{1 + \lambda \cdot \mu_A(g)}} \qquad (5.14)$$

$$= \frac{1 - \mu_A(g)}{1 + 3 \cdot \lambda \cdot \mu_A(g) + \lambda^2 \mu_A(g) + \mu_A(g)}$$

The hesitation degree is calculated as $\pi_{mn} = 1 - \mu_A^{IFS}(g) - v_A^{IFS}(g)$.

λ is calculated using the *IF* entropy. The *IF* entropy is calculated as

$$IE(A) = \frac{1}{N} \sum_{j=1}^{N} \sum_{i=1}^{N} \pi_A(g_{ij}) \cdot e^{1 - \pi_A(g_{ij})}$$

The optimum value of λ is calculated as

$$\lambda_{opt} = \max_{\lambda}(IE(A; \lambda))$$

Among all the entropic values for different values of λ, the λ value that corresponds to the maximum entropy is selected. With this λ value, the membership and the hesitation degrees are calculated.

Then fuzzy hedge is applied on the *IF* image which is given as

$$\mu_{new}^{IFS}(g) = \left(\mu_A^{IFS}(g)\right)^{1.25}$$

Contrast stretching is applied on the intuitionistic image using the INT operator.

The intensifier operation is written as

$$\mu^{enh}(g) = \begin{array}{ll} 2 \cdot \left[\mu_{IFS}^{new}(g)\right]^2 & \text{if } 0 < \mu_{IFS}^{new}(g) \le 0.5 \\[2mm] 1 - 2\left[1 - \mu_{IFS}^{new}(g)\right]^2 & \text{if } 0.5 \le \mu_{IFS}^{new}(g) \le 1 \end{array}$$

where μ^{enh} is the enhanced *IF* image.

5.4.4 Contrast Enhancement by Chaira (Method IV)

This method is suggested by Chaira [7] using Chaira's *IF* generator. The image is fuzzified with the membership function $\mu_A(g)$ using Equation 5.1.

Based on the fuzzy set, the membership degree of the *IF* image is computed from Chaira's *IF* generator as

$$\mu_A^{IFS}(g) = 1 - \frac{1-\mu_A(g)}{1+(e^\lambda-1)\mu_A(g)} \tag{5.15}$$

with $\lambda > 0$.

Using Chaira's fuzzy negation, $\varphi(x) = (1-x)/1 + (e^\lambda - 1)x$, $\lambda > 0$, the non-membership degree of the *IF* image is computed as

$$v_A^{IFS}(g;\lambda) = \varphi\left(\mu_A^{IFS}(g;\lambda)\right)$$

or

$$v_A^{IFS}(g) = \frac{1-\mu_A^{IFS}(g)}{1+(e^{\lambda+1}-1)\mu_A^{IFS}(g)} \tag{5.16}$$

λ in Equation 5.16 in the denominator is changed to $\lambda + 1$, implying that $1+(e^{\lambda 1}-1)\mu_A^{IFS}(g)$ is changed to $1+(e^{\lambda+1}-1)\mu_A^{IFS}(g)$. With the change in λ, the non-membership degree, $v_A^{IFS}(g)$, will change but will still follow the condition $v_A^{IFS}(g) \le 1-\mu_A^{IFS}(g)$. This is done to obtain a better contrast-enhanced image. So,

$$
\begin{aligned}
v_A^{IFS}(g;\lambda) &= \frac{1-\left(1-\dfrac{1-\mu_A(g)}{1+(e^\lambda-1)\mu_A(g)}\right)}{1+(e^{\lambda+1}-1)\left(1-\dfrac{1-\mu_A(g)}{1+(e^\lambda-1)\mu_A(g)}\right)} \\[2mm]
&= \frac{1-\mu_A(g)}{1+(e^\lambda-1)\mu_A(g)+(e^{\lambda+1}-1)\mu_A(g)e^\lambda} \tag{5.17}\\[2mm]
&= \frac{1-\mu_A(g)}{1+(e^{2\lambda+1}-1)\mu_A(g)}
\end{aligned}
$$

The hesitation degree is computed as

$$\pi_A^{IFS}(g;\lambda) = 1 - \mu_A^{IFS}(g;\lambda) - v_A^{IFS}(g;\lambda) \tag{5.18}$$

To obtain the optimum value of λ, the *IF* entropy is used.

The *IF* entropy is calculated as

$$IE(A) = \frac{1}{N}\sum_{j=1}^{N}\sum_{i=1}^{N}\pi_A(g_{ij})\cdot e^{1-\pi_{IFS}(g_{ij})}$$

The optimum value of λ is calculated as

$$\lambda_{opt} = \max_{\lambda}(IE(A;\lambda))$$

The λ value that corresponds to the maximum entropy is selected. With this λ value, the membership and the hesitation degrees are calculated.

With the λ value, the membership function in Equation 5.15 and the non-membership function in Equation 5.17 are computed and an *IF* image is obtained.

Then a contrast intensifier is applied to the *IF* image which is written as

$$\mu^{enh}(g) = \begin{array}{ll} 2\left[\mu_A^{IFS}(g)\right]^2 & \text{if } \mu_A^{IFS}(g) \leq 0.5 \\ 1 - 2\left[1 - \mu_A^{IFS}(g)\right]^2 & \text{if } 0.5 < \mu_A^{IFS}(g) \leq 1 \end{array}$$

This enhanced image is the contrast-enhanced image.

Example 5.3

Two examples of medical images using *IF* methods by Chaira (methods III and IV) are shown in Figures 5.5 and 5.6 to illustrate the efficacy of the methods.

5.4.5 Hesitancy Histogram Equalization

In this method, a hesitant histogram is generated, and then using histogram equalization, image enhancement is obtained.

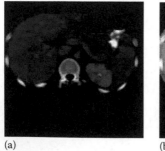 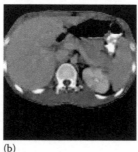 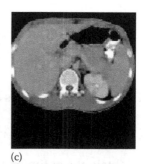

(a)   (b)   (c)

FIGURE 5.5
(a) CT scan brain (CT-1) image, (b) *IF* enhancement by Chaira (method III) and (c) the *IF* method by Chaira (method IV). (Modified from Chaira, T., Construction of intuitionistic fuzzy contrast enhanced medical images, in *Proc. of IEEE International Conference on Human Computer Interaction*, IIT Kharagpur, India, 2012; Chaira, T., *J. Intell. Fuzzy Syst.*, 2013.)

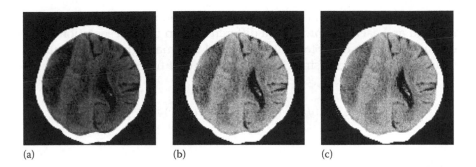

(a)   (b)   (c)

FIGURE 5.6
(a) CT scan brain (CT-2) image, (b) *IF* enhancement by Chaira (method III) and (c) the *IF* method by Chaira (method IV). (Modified from Chaira, T., Construction of intuitionistic fuzzy contrast enhanced medical images, in *Proc. of IEEE International Conference on Human Computer Interaction*, IIT Kharagpur, India, 2012; Chaira, T., *J. Intell. Fuzzy Syst.*, 2013.)

Method V: Vlachos and Sergiadis [18] suggested hesitancy histogram equalization for image enhancement. As has been described earlier, an *IF* image is written as

$$A^{IFS} = \left\{ x, \mu_A(x), \nu_A(x) \middle| x \in X \right\}$$

where $\mu_A(x)$ and $v_A(x)$ are the membership and non-membership functions in the interval [0, 1], respectively.

Using the concept of the fuzzy histogram, an *IF* histogram is constructed using an *IF* number. A symmetrical triangular *F*-number is used to construct the fuzzy histogram and is represented as

$$\mu_g^F(x) = \max\left(0, 1 - \frac{|x - g|}{p} \right)$$

where p is a real number that controls the shape of the *F*-number. Using the *IF* generator, the *IF* membership function is written as

$$\mu_g(x) = 1 - \left(1 - \max\left(0, 1 - \frac{|x - g|}{p} \right) \right)^{\lambda - 1}$$

and the non-membership function is written as

$$\nu_g(x) = \left(1 - \max\left(0, 1 - \frac{|x - g|}{p} \right) \right)^{\lambda(\lambda - 1)}$$

By varying the λ value, various membership and non-membership values are obtained and each one provides a different notion of representation of grey level around 'g'. In order to optimally model the image grey levels, optimum value of λ is required. Using *IF* entropy, λ_{opt} is calculated as

$$IFE(A) = \sum_{g=0}^{L-1} h_A(g)E_g(A;\lambda)$$

where

$$E_g(A;g) = \sum_{k=0}^{L-1} h_A(k)\left(1-\mu_g^A(k;\lambda)-v_g^A(k;\lambda)\right)$$

$$= \sum_{g=0}^{L-1} h_A(g)\sum_{k=0}^{L-1} h_A(k)\left(\left(1-\max\left(0,1-\frac{|k-g|}{p}\right)\right)^{\lambda-1}-\left(1-\max\left(0,1-\frac{|k-g|}{p}\right)\right)^{\lambda.(\lambda-1)}\right)$$

$$(5.19)$$

h_A is the crisp histogram of the image.

The optimal value of λ corresponds to the maximum entropy value. Based on the lower and upper membership functions, the lower (minimum) *IF* histogram is written as follows.

Then lower *IF* histogram is

$$h_A^L(g) = \left\{(i,j),\mu_g^A(g_{ij},\lambda_{opt})\right\}$$

and upper (maximum) *IF* histogram is

$$h_A^U(g) = \left\{(i,j),1-v_g^A(g_{ij},\lambda_{opt})\right\}$$

The hesitancy histogram of an image is then given as

$$h_A^H(g) = h_A^U(g)-h_A^L(g), \quad g = 0,1,2,\ldots,L-1$$

The normalized hesitancy histogram is then computed as

$$h_A^H(g) = \frac{h_A^U(g)-h_A^L(g)}{\sum_{k=0}^{L-1} h_A^U(k)-h_A^L(k)}$$

$$(5.20)$$

The interval $[h_A^U(g), h_A^L(g)]$ can be interpreted as the possible frequency of occurrence of the intensity level around 'g' with the lower bound representing the minimum possible frequency of occurrence of the grey level and the upper bound representing the maximum frequency of occurrence of the grey level. Thus, hesitancy histogram is the length of the interval. Now for hesitance histogram equalization, similar to the conventional histogram equalization method, the cumulative density function of the hesitancy histogram is used for grey-level transformation so that the resulting image will possess a uniform hesitancy histogram:

$$g' = (L-1) \sum_{k=0}^{g} h_A^H(g) \tag{5.21}$$

where g and g' are the original and transformed grey levels of the image, respectively. A contrast-enhanced image is formed.

5.5 Image Enhancement Using Type II Fuzzy Set

In this section, image enhancement using Type II fuzzy set is discussed. It considers the membership function in ordinary fuzzy set as fuzzy or vague, and so the membership function lies in an interval range with upper and lower membership levels. Thus, Type II fuzzy set represents the uncertainty in a different and better way than type I fuzzy set, and so better enhancement results may be expected. There is very little research on medical image enhancement using Type II fuzzy set, and these are discussed in this section.

5.5.1 Type II Fuzzy Enhancement (Method I)

Chaira suggested a Type II fuzzy enhancement method using fuzzy t-conorm by Chaira discussed in Chapter 3. Though there are many t-conorms in the literature, the t-conorms that are algebraic in nature perform better. The reason behind this is that the operators that belong to a conditional class may not lead to realization of perfect t-norm and t-conorm though they are computationally simple. The operators in the algebraic class that do not contain the min and max operators reveal the actual value.

The original image is initially fuzzified with the membership function $\mu_A(g)$ using Equation 5.1. Then using Type II fuzzy set, two levels are computed:

$$\mu^{upper}(g) = [\mu(g)]^{\alpha}$$

$$\mu^{lower}(g) = [\mu(g)]^{1/\alpha}, \quad 0 < \alpha \leq 1 \tag{5.22}$$

A new membership function is computed using fuzzy *t*-conorm by Chaira, which is algebraic in nature and is computed as

$$\mu^{enh}(g) = \frac{\mu^{upper}(g) + \mu^{lower}(g) + \lambda \cdot \mu^{upper}(g) \cdot \mu^{lower}(g)}{(1+\lambda) \cdot \mu^{upper}(g) \cdot \mu^{lower}(g) + 1}$$

This is obtained from

$$C^*(x,y) = \frac{x + y + \lambda xy}{(1+\lambda)xy + 1}$$

$\mu^{upper}(g)$ and $\mu^{lower}(g)$ are the upper and lower membership functions of the Type II fuzzy set, respectively.

$\lambda = im\_avg$, where im_avg is the average of the image. The new image with the new membership function so formed is the enhanced image.

To compute the value of α, fuzzy linguistic hedge $0 \leq \alpha \leq 1$ is used to generate the lower and upper membership functions from a type I fuzzy membership function. Parameter α is usually determined heuristically to satisfy the requirement $0 \leq \alpha \leq 1$. In this method, $\alpha = 0.75$ is used. The upper and lower ranges of the Type II fuzzy membership function are calculated with $\alpha = 0.75$ in Equation 5.18.

5.5.2 Enhancement Using Hamacher *t*-Conorm

A similar procedure as described earlier is followed using Hamacher *t*-conorm [11]. The new membership function is computed using Hamacher *t*-conorm as

$$\mu^{enh}(g) = \frac{\mu^{upper}(g) + \mu^{lower}(g) + (\lambda - 2) \cdot \mu^{upper}(g) \cdot \mu^{lower}(g)}{1 - (1-\lambda) \cdot \mu^{upper}(g) \cdot \mu^{lower}(g)}$$

and $\lambda = im\_avg$, where im_avg is the average of the image. As in the previous method, the upper and lower ranges of the Type II fuzzy membership function are calculated with $\alpha = 0.8$. The new image so formed is the enhanced image [5].

Enhancement using other *t*-conorms may also be used using the same procedure.

5.5.3 Enhancement Using Type II Fuzzy Set

This type of enhancement was suggested by Ensafi and Tizhoosh [8]. It considers Type II fuzzy set to enhance the images. The image is divided into several windows, and for each window, the subimage is fuzzified using Equation 5.1.

The upper and lower membership values are computed as

$$\mu^{upper}(g) = [\mu(g)]^{0.5}$$
$$\mu^{lower}(g) = [\mu(g)]^{2}$$

where $\mu(g)$ is the fuzzified image.

From the two levels, the new membership function is computed as

$$\mu_{new}(g) = \mu^{lower}(g) \cdot \alpha + \mu^{upper}(g) \cdot (1-\alpha), \quad 0 < \alpha < 1$$

and

$$\alpha = \frac{g_{mean}}{L}$$

where

g_{mean} is the mean of the grey levels of the window

L is the number of grey levels

This is done for all the windows. The new image obtained is the enhanced image.

Example 5.4

Three results on abdominal tumour and ovarian cyst will illustrate the effect of the Type II fuzzy method in image enhancement. The results on different types of images are shown in Figures 5.7 through 5.9.

For performance evaluation, the linear index of fuzziness is computed for the original and the enhanced images. It determines the vagueness in the image, and enhancement takes place when there is a decrease in fuzziness, that is, if the index of fuzziness is less than the original image, the enhanced image is better. It gives at least some useful information about the efficacy of the methods. The linear index of fuzziness is defined as

$$L.I. = \frac{2}{M \times N} \sum_{i=0}^{M-1} \sum_{j=0}^{M-1} \min\left(\mu^{enh}(g_{ij}),\ 1-\mu^{enh}(g_{ij})\right) \qquad (5.23)$$

where

$\mu^{enh}(g_{ij})$ is the membership value of the enhanced image

$M \times N$ is the size of the image

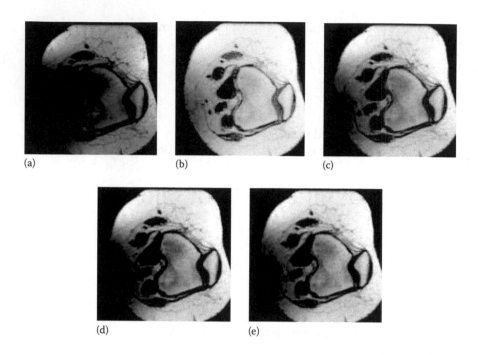

FIGURE 5.7
(a) Knee patella image, (b) enhanced image using Dombi *t*-conorm, (c) enhanced image (Ensafi), (d) enhanced image with Chaira *t*-conorm and (e) enhanced image with Hamacher *t*-conorm. (Modified from Chaira, T., Contrast enhancement of medical images using Type II fuzzy set, in *Proc. of IEEE on National Conference on Communication*, IIT Delhi, India, 1–5, 2013.)

Actually the medical images are poorly illuminated, so the background is not bright and the inner structures are not clear. Enhancement should be such that the background is brighter and the structures are highlighted to make the inner structures clearly visible, which is obtained in the proposed Type II fuzzy method.

5.6 Introduction to MATLAB®

In this section, few MATLAB® examples are given. Before writing the programs, a brief overview of MATLAB is given that will help the students to understand the programs very easily. In many cases, the students sometimes find it difficult to implement the image processing algorithms in MATLAB. In view of this, a brief note on MATLAB, followed by common MATLAB commands and finally the MATLAB codes of the various algorithms, is presented. MATLAB is a Matrix Laboratory; it was invented in the late 1970s by Cleve Moler. It is a numerical computing method for writing programs.

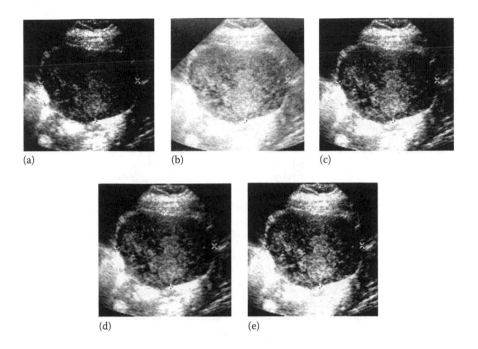

(a)   (b)   (c)

(d)   (e)

FIGURE 5.8
(a) Abdominal tumour, (b) enhanced image using Dombi *t*-conorm, (c) enhanced image (Ensafi), (d) enhanced image with Chaira *t*-conorm and (e) enhanced image with Hamacher *t*-conorm. (Modified from Chaira, T., Contrast enhancement of medical images using Type II fuzzy set, in *Proc. of IEEE on National Conference on Communication*, IIT Delhi, India, 1–5, 2013.)

It was created by MathWorks and provides easy matrix (2D) or vector (a 1D matrix) calculations; plotting of functions, data and graphs; implementing algorithms; and interfacing with programs in other languages. To mention a few, it is also used in education such as in teaching linear algebra, numerical analysis, differential equation, fuzzy mathematics, signal processing and neural network and is very much popular among scientists involved with image processing. MATLAB consists of many commands, which directly compute different matrix operations such as multiplication, transpose, eigenvalues and other functions very quickly. There are many examples available in the MATLAB Toolbox. The best way to learn MATLAB is to work through some examples at the computer. MATLAB commands and MATLAB help are available online. To mention a few, toolboxes on signal processing, image processing, neural network, fuzzy mathematics and many others are available.

Image is a 2D matrix, and thus using MATLAB program becomes very easy instead of writing in C. It performs intensive tasks faster than that of the traditional programming languages such as C, C++ and FORTRAN. MATLAB also includes 'for', 'while' loops and 'if-else' commands as in C.

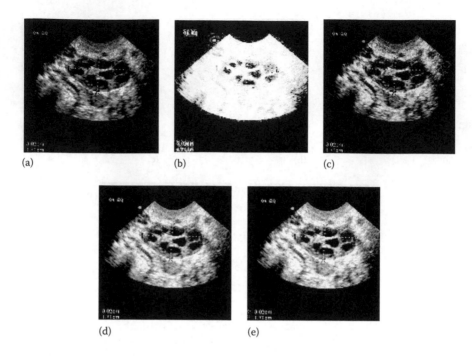

FIGURE 5.9
(a) Polycyst, (b) enhanced image using Dombi *t*-conorm, (c) enhanced image (Ensafi), (d) enhanced image with Chaira *t*-conorm and (e) enhanced image with Hamacher *t*-conorm. (Modified from Chaira, T., Contrast enhancement of medical images using Type II fuzzy set, in *Proc. of IEEE on National Conference on Communication*, IIT Delhi, India, 1–5, 2013.)

MATLAB program and files always have filenames ending with `'filename.m'`.

Some of the basic matrix commands used in image processing are as follows:

```
To display a matrix
a = [1 3; 2 4]
b = a*a (element by element multiplication)
b = [1 9; 4 16]
Addition of two matrices - a + b
Inverse of a matrix - inv(a)
zeros (1, 3)
ones (1, 3)
```

Some of the MATLAB commands used in image processing are as follows:

```
1. Reading an image A.
X = imread(A)
2. Displaying an image
imshow(A)
```

3. Histogram of an image A
imhist(A)
4. Histogram equalization
histeq(A)
5. Cropping an image, A, from a coordinate point of dimension
  'dim' – A small portion from the image is cropped.
imcrop (A,[x y [dim dim]])
6. Rotating an image by 30°
imrotate (A, 30)
7. Convert to unsigned 8-bit integer. The command (uint8)
  converts a matrix into an unsigned 8-bit integer. B can be
  any numeric object such as Double. The elements of an UINT8
  range from 0 to 255. The result for any element of B outside
  this range is not defined and they are either assigned to 0
  or 255. It is written as
C = uint8(B)
8. Finding the edges of the image. There are many edge
  detection operators such as Roberts, Sobel, Prewitt,
  Laplacian of Gaussian(LoG), Canny. An example using Prewitt
  edge detector is as follows:
A = imread('cameraman.tif');
EIM = edge(I,'prewitt', thresh);
imshow(A), figure, imshow(EIM)
9. To convert an image to a binary image
BIM = graythresh(A)
10. Converting a RGB color image to gray image
rgb2gray(a)
11. Performing morphological operators on the binary image.
  There are many operations such as erode, dilate, thin, and
  open remove. An example of "thin" an image is shown.
A = imread('cameraman.tif');
imshow(A)
IM1 = bwmorph(BW1,'dilate');
figure, imshow(IM1)

5.6.1 Examples Using MATLAB

A MATLAB code for image enhancement is given which will be beneficial to
the readers to implement the method.

1. *Type II fuzzy image enhancement using Hamacher t-conorm*

```
image=imread('knee.jpg');
b=rgb2gray(image);
dim=140;
c=imcrop(b,[1 1 dim-1 dim-1]);
d=double(c);
mx=max(max(d));
mn=min(min(d));
```

```
img_mem=(d-mn)./(mx-mn);
f0 = [];
alpha1=0.7;
     im_avg=mean2(img11);        % find the mean of the image
     alpha2 = im_avg;
     m_high=img_mem.^alpha1;      % upper membership level
     m_low =img_mem.^(1/alpha1);  % lower membership level
% new membership function
mem_hama =
          (m_low+m_high+m_low.*m_high*(alpha2-2))./(1+m_low.*m_
          high*(alpha2-1));
  figure, imshow(mem_hama)        % Type II fuzzy enhanced image
```

2. *A program to produce an intuitionistic fuzzy–enhanced image*

```
image1=imread('knee.jpg');
a1=rgb2gray(image1);
dim=140;
img1=imcrop(a1,[2 1 dim-1 dim-1]);
img=double(img1); mx=max(max(img)); mn=min(min(img));
mem1=(img-mn)./(mx-mn);
mun=[];
% computing Intuitionistic fuzzy membership function
%% finding optimum value of con
 for con=1:0.1:10
    mem   =1-(1-mem1).^con;
    nonmem=(1-mem1).^(con*(con+1));
    hes=1-mem-nonmem;
    u=0.0;
    for i=1:dim
       for j=1:dim
    ent=(2*mem(i,j)*nonmem(i,j)+hes(i,j)^2)/(hes(i,j)^2+
       mem(i,j)^2+ nonmem(i,j)^2) +u;
       u=ent;
       end
    end
  lin=ent;
  lin_ind=[mun;lin];
  mun=lin_ind;
  lin_ind;
end
lin_ind;
  l=max(lin_ind);
  [con]=find(l==lin_ind);
  fincon= con*0.1+1;
% Intuitionistic fuzzy membership and non-membership functions
newmem =1-(1-mem1).^fincon;
newnonmem =(1-mem1).^(fincon*(fincon+1));
newhes    =1-newmem-newnonmem;
```

```
% Atanassov't operator to convert intuitionistic fuzzy
   image to fuzzy image
v1=0.0;v2=0.0;
for i=1:dim
    for j=1:dim
    num= (newhes(i,j)*(1-2*newmem(i,j)))+ v1;
    v1=num;
    denom=(newhes(i,j)^2)+ v2;
    v2=denom;
    end
    end
alpha_opt1 =num/(2*denom);
if alpha_opt1>0 & alpha_opt1<1
alpha_opt=alpha_opt1;
end
D_alpha_opt=(1-alpha_opt)*newmem + (1-newnonmem)*alpha_opt;
enh_im=uint8(255*D_alpha_opt);
figure,imshow(enh_im); % final enhanced image
```

5.7 Summary

In this chapter, image enhancement using fuzzy, *IF* and Type II fuzzy set theories is suggested. Enhancement is the preprocessing of images to enhance or highlight the image structures and suppress unwanted information in the image. Fuzzy enhancement does provide better results, but in some cases, *IF* and Type II fuzzy enhancement methods provide better results. This may be due to the fact that these advanced fuzzy sets consider either more number of uncertainties or different types of uncertainty. Also, MATLAB codes of various types of image enhancement schemes proposed by different authors are discussed.

References

1. Acharya, T. and Ray, A.K., *Image Processing: Principles and Application*, John Wiley and Sons, 2005.
2. Atanassov, K.T., *Intuitionistic Fuzzy Sets: Theory and Applications*, Series in Fuzziness and Soft Computing, Physica-Verlag, Heidelberg, Germany, 1999.
3. Ban, A.I., *Intuitionistic Fuzzy Measures: Theory and Applications*, Nova Science Publishers, New York, 2006.

4. Burillo, P. and Bustince, H., Entropy on intuitionistic fuzzy sets and on interval valued, *Fuzzy Sets and Systems*, 78, 305–316, 1996.
5. Chaira, T., Contrast enhancement of medical images using type II fuzzy set, in *Proc. of IEEE National Conference on Communication*, IIT Delhi, India, 1–5, 2013.
6. Chaira, T., Construction of intuitionistic fuzzy contrast enhanced medical images, in *Proc. of IEEE International Conference on Human Computer Interaction*, 1–5, IIT Kharagpur, India, 2012.
7. Chaira, T., Enhancement of medical images using Atanassov's intuitionistic fuzzy domain using an alternative intuitionistic fuzzy generator with application to medical image segmentation, *Journal of Intelligent and Fuzzy Systems*, 27(3), 1347–1359, 2014.
8. Ensafi, P. and Tizhoosh, H.R., Type II fuzzy image enhancement, in *Lecture Notes in Computer Sciences*, Vol. 3656, M. Kamel and A. Campilho (Eds.), Springer, Berlin, Germany, pp. 159–166, 2005.
9. Friedman, M., Schneider, M., and Kandel, A., Properties of fuzzy expected value and fuzzy expected interval in fuzzy environment, *Fuzzy Sets and Systems*, 28, 55–68, 1988.
10. Gonzales, R.C. and Woods, R.E., *Digital Image Processing*, Addison-Wesley Publishing Group, Reading, MA, 1992.
11. Hamacher, H., *Über logische Aggregation nicht-binär explizierter Entscheidnungskriterien*, R.G. Fisher Verlag, Frankfurt, Germany, 1978.
12. Handmandlu, M., Jha, D., and Sharma, R., Color image enhancement using fuzzy intensification, *Pattern Recognition Letters*, 24, 81–87, 2004.
13. Hassanien, A.E. and Badr, A., A comparative study on digital mammography enhancement algorithm based on fuzzy set theory, *Studies in Information and Control*, 12(1), 21–31, 2003.
14. Schneider, M. and Craig, M., On the use of fuzzy sets in histogram equalization, *Fuzzy Sets and Systems*, 45, 271–278, 1992.
15. Sonka, M. et al., *Image Processing Analysis and Computing Vision*, Brooks/Cole, Pacific Grove, CA, 2001.
16. Tizhoosh, H.R. and Fochem, M., Fuzzy histogram hyperbolization for image enhancement, in *Proc. of EUFIT'95*, Vol. 3, 1695–1698, Aachen, Germany, 1995.
17. Vlachos, I.K. and Sergiadis, G.D., Role of entropy in intuitionistic fuzzy contrast enhancement, in *Lecture Notes in Artificial Intelligence*, Vol. 4529, P. Melin et al. (Eds) Springer-Verlag Berlin Heidelberg, pp. 104–111, 2007.
18. Vlachos, I.K. and Sergiadis, G.D., Hesitancy histogram equalization, in *Proc. of FUZZ-IEEE*, London, U.K., pp. 1–6, 2007.
19. Vlachos, I.K. and Sergiadis, G.D., A two-dimensional entropic approach to intuitionistic fuzzy contrast enhancement, in *Proc. of IEEE WILF*, Italy, pp. 321–327, 2007.
20. Vlachos, I.K. and Sergiadis, G.D., Intuitionistic fuzzy image processing, in *Soft Computing in Image Processing: Recent Advances*, Studies in Fuzziness and Soft Computing, Vol. 210, Springer, Heidelberg, Germany, pp. 385–416, 2006.

6

Thresholding of Medical Images

6.1 Introduction

Segmentation is a fundamental building block in image processing analysis. It is the first stage of analysing an image. It partitions the image into disjoint regions, which correlate strongly with objects or features of interest where pixels of similar attributes are grouped together. Segmentation techniques may be non-contextual or contextual. In the non-contextual technique, features of an image are not taken into account and the pixels are grouped together on the basis of a global attribute, that is, pixel intensities. It does not take into account the relative location of the pixels. The contextual technique takes into account the closeness of the pixels in an object. That is, it exploits the relationship between the image features. Thresholding is a simple and non-contextual technique. It is computationally inexpensive and fast. Thresholding is a segmentation technique that classifies the pixels into two categories: those pixels that fall below the threshold and those that fall above the threshold. It involves analysing the histogram. A threshold may be global or local. Global threshold selects a threshold for the entire image. This method does not work if there is uneven illumination, and in that case, local threshold works well. In local thresholds, thresholds are obtained from each subregion of an image, thereby adapting to local variations.

A complete segmentation of an image R is a finite set of regions R_1, R_2, R_3, ..., R_N, such that

$$R = \bigcup_{i=1}^{N} R_i \quad \text{and} \quad R_i \cap R_j = \phi, \quad i \neq j$$

Thresholding is a transformation of an input image A into a segmented output image B as follows:

$$b_{ij} = 1 \quad \text{for } a_{ij} \geq T$$
$$= 0 \quad \text{for } a_{ij} \leq T$$

where
$\quad T$ is the threshold
$\quad b_{ij} = 1$, for the image pixels that belong to the object class
$\quad b_{ij} = 0$, for the image pixels that belong to the background class

However, the selection of threshold is very crucial in segmenting an image. If the threshold is not selected properly, proper segmentation is not obtained. A threshold may be global or local. In global thresholding, the image is considered as a whole and the threshold value is constant throughout the image. Depending on the modality of the histogram, threshold levels may be single, double or multiple. In single threshold, the image contains an object against a background. The histogram of such type of image contains two peaks – one for the object and the other for the background – and the threshold value lies in the valley of the two peaks. In double threshold, the image consists of two objects of different grey levels and the histogram contains three peaks – two for the objects and one for the background. The thresholds lie in the two valleys of the three peaks.

In local threshold, the threshold varies throughout the image. In some images, different objects of different grey levels may be presented, and in that, global threshold will not work. The image is divided into subregions, and the threshold is computed for each region. This is useful in processing medical images where the grey levels are not uniformly distributed.

6.2 Threshold Detection Methods

There are six methods of thresholding, namely, (1) global thresholding, (2) iterative thresholding, (3) optimal thresholding, (4) local thresholding, (5) the Chow and Kaneko approach for adaptive threshold and (6) multispectral thresholding.

6.2.1 Global Thresholding

1. *P-tile thresholding*: On a printed sheet text, if the area covered by the text is $1/p$ of the area of the image, then the threshold, T, chosen is such that $1/p$ of the image area has grey level less than the threshold, T, and the rest of the values are larger than the threshold, T.

2. If an image consists of distinct objects and background, the histogram will contain peaks and valleys. If the histogram is bimodal, the threshold is selected at the valley of the two peaks, that is, at the minima. If the histogram is multimodal, more thresholds may be determined at minima between any two maxima.

3. Another technique that constructs a grey-level histogram consists of border pixels and the histogram thus making the histogram unimodal. The threshold is chosen at the peak of the histogram, and the peak corresponds to the grey levels bordering between the object and the background.

4. For obtaining distinct peaks and valleys, one can weigh the histogram by suppressing the pixels with a high gradient. In such cases, the histogram will contain the grey levels that belong mostly to the object and the background and may not consider the border pixels having a high pixel gradient.

6.2.2 Iterative Thresholding

In this method, the threshold is selected iteratively:

1. Select an initial estimate of the threshold T, which may be the mean.
2. Using the threshold or the mean, partition the image into regions as R_1 and R_2.
3. Calculate the mean of the intensities of the two regions as μ_1 and μ_2.
4. Compute the new threshold as $T = (\mu_1 + \mu_2)/2$.
5. Repeat steps 2–4 until the mean values μ_1 and μ_2 do not change in the successive iterations.

6.2.3 Optimal Thresholding

In many images, histogram peaks are not clearly separable and it happens when the object is flat with no colour variation and no discernible texture and colour. Peaks may overlap and choosing a threshold in overlapping peaks will not classify the pixels in correct regions. So, many pixels will be incorrectly classified. Optimal thresholding uses a criterion function that yields a measure of separation between two regions. A criterion function may be calculated in many ways that may be Shannon's entropy, cross entropy, divergence or any other measure. There are many thresholding techniques that use the optimal method to select a threshold [19,20].

6.2.4 Locally Adaptive Thresholding

Till now, global thresholding methods are discussed. Global threshold selects only one threshold for the entire image. But in some images, especially in medical images where there is a variation in grey-level intensity throughout the image, global threshold does not work better. Local or window-based thresholding methods may be very much effective where the image is divided into several windows or regions and there is one threshold (or one set of thresholds) for each window. The threshold value depends on the local statistics of the region such as the variance and mean of the image. It may be termed as regional thresholding. It works well when the intensities are not uniform and multiple objects are present with different grey levels. Such types of images are the real-time images, for example, medical images where the images are not uniform due to improper illumination or lightning condition or any other. One possible way for local threshold [14] is replacing each pixel by the mean and standard deviation of its local neighbourhood of size $b \times b$, which may be calculated as

$$T(i, j) = m(i, j) + k \cdot \sigma(i, j)$$

where m, σ and k are the local mean, variance and bias setting, respectively.

Rodriguez [18] suggested a non-fuzzy/crisp method on segmenting medical images using windowed thresholding based on Otsu's method [15], where the image in the object class (C_0) and background class (C_1) is separated using a threshold t. Between-class variance σ_b^2 and within-class variance σ_w^2 are calculated for the image, and the optimal threshold is selected by maximizing the ratio of between-class variance to within-class variance with respect to the threshold t, that is, maximizing the separability of the histogram:

$$\eta = \frac{\sigma_b^2}{\sigma_w^2}$$

This approach works better than any other image thresholding methods. It performs better where the contrast of the background and object region is low but sensitive to noise. So, a variation of Otsu's method was suggested where the image is divided into windows of size $(1/2) \times (1/4)$ of the image and filtered the image using a Gaussian filter of $\sigma = 0.5$. The standard deviation is calculated for each window:

$$\text{S.D.} = \sqrt{\frac{1}{w^2} \sum_{k=1-M}^{i+M} \sum_{l=j-M}^{j+M} \left(f(k,l) - \mu(i,j)\right)^2}$$

where
$M = (W - 1)/2$, W denotes the width of the window
$$\mu(i, j) = 1/W^2 \sum_{k=1-M}^{i+M} \sum_{l=j-M}^{j+M} f(k,l)$$

The final threshold of each window is Otsu's threshold minus the standard deviation.

So, $\text{Threshold}_{window} = \text{Threshold}_{Otsu} - \text{S.D.}_{window}$

This process is carried for all windows, and the thresholded images are combined to form a final image.

Example 6.1

Two examples of medical images utilizing window-based thresholding are shown in Figure 6.1.

6.2.5 Locally Adaptive and Optimal Thresholding

A combination of adaptive and optimal thresholding gives better result. This method determines optimal grey-level thresholding parameters in local subregions for which local histograms are constructed.

6.2.5.1 *Chow and Kaneko Method*

According to Chow and Kaneko, the image is divided into an array of overlapping subregions and the grey-level histogram is constructed for each subregion.

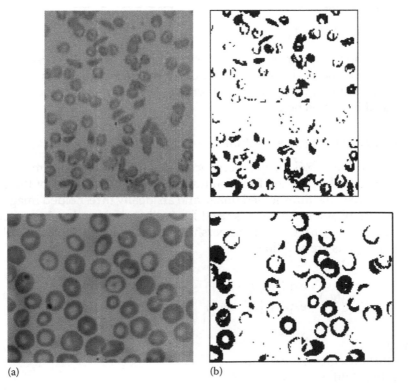

(a)   (b)

FIGURE 6.1
(a) Blood vessel image and (b) local thresholding method by Rodriguez.

Optimal threshold is computed for each region. The threshold for each pixel is computed by interpolating the results of the subimages. This method gives a better result, but computational complexity is more in real-time applications.

An alternative method to find local threshold is to look into the local statistics of each pixel around its neighbourhood such as the mean or median or mean of the difference of minimum and maximum values in the subimage.

6.2.5.2 Multispectral Thresholding

This type of thresholding is used in remotely sensed images where there are many spectral bands. One simple way is to obtain a threshold for each band and then combine to obtain a thresholded image.

Though this book mainly discusses medical image processing using advanced fuzzy set theories, a brief overview on fuzzy thresholding on medical images is discussed. The methods are not discussed in detail as these are already presented in the author's other book on fuzzy image processing. Approaches related to intuitionistic fuzzy set and Type II fuzzy set will be discussed in detail with examples on medical images in later sections.

6.3 Fuzzy Methods

In this section, fuzzy methods are discussed briefly so that the readers may have a clear knowledge on the application of fuzzy sets in thresholding. There are many fuzzy methods in literature [10–13,16], and few of them are explained in this section.

6.3.1 Fuzzy Divergence Method

Chaira and Ray's [5,6] fuzzy divergence method is used for optimal threshold selection. The criterion function involves the minimization of the divergence between the thresholded image and an ideally thresholded image. The gamma membership function is used for finding the membership values of the pixels in an image. The membership functions for the object and background regions are as follows:

$$\mu_A(a_{ij}) = \frac{\exp(-c \cdot |a_{ij} - m_0|)}{\exp(-c \cdot |a_{ij} - m_1|)} \quad \begin{array}{l} \text{if } a_{ij} \leq t, \text{ for object} \\ \text{if } a_{ij} > t, \text{ for background} \end{array} \tag{6.1}$$

where
    t is the any chosen threshold
    a_{ij} is the (i, j)th pixel in image A
    constant $c = 1/(f_{max} - f_{min})$
    m_0 and m_1 are the means of object and background regions, respectively

$$m_0 = \frac{\sum_{f=0}^{t} f \cdot count(f)}{\sum_{f=0}^{t} count(f)}, \quad m_1 = \frac{\sum_{f=t+1}^{L-1} f \cdot count(f)}{\sum_{f=t+1}^{L-1} count(f)}$$

The optimal threshold is selected by minimizing fuzzy divergence. Fuzzy divergence between two images A and B is written as

$$\sum_{i=0}^{M-1}\sum_{j=0}^{M-1}\left(2-(1-\mu_A(a_{ij})+\mu_B(b_{ij}))e^{\mu_A(a_{ij})-\mu_B(b_{ij})} - (1-\mu_B(b_{ij})+\mu_A(a_{ij}))e^{\mu_B(b_{ij})-\mu_A(a_{ij})}\right)$$

For thresholding purposes, fuzzy divergence between the thresholded image and the ideally thresholded image, $\mu_B(b_{ij}) = 1$, is written as

$$D(A,B) = \sum_{i=0}^{M-1}\sum_{j=0}^{M-1}\left(2-(2-\mu_A a_{ij}) \cdot e^{\mu_A a_{ij}-1} - \mu_A a_{ij} \cdot e^{1-\mu_A a_{ij}}\right) \qquad (6.2)$$

Divergence is calculated for all threshold grey levels, and the grey level corresponding to the minimum divergence is selected as the optimal threshold.

Image thresholding using *measures of fuzziness* and *fuzzy entropy* follows the same procedure as fuzzy divergence where the measures of fuzziness or entropy are minimized.

6.3.2 Fuzzy Geometry Method

Rosenfeld [17] introduced the concept of fuzzy geometry of image subsets such as area, perimeter and compactness. Pal and Rosenfeld [16] used the concept of fuzzy compactness to obtain an optimal threshold and standard Zadeh's S-function for finding the membership values of the pixels in an image. The optimal threshold is selected by minimizing the fuzzy compactness. Compactness for a threshold t_i is given as

$$Compactness(\mu) = \frac{area(\mu)}{per(\mu)^2}$$

where area and perimeter corresponding to threshold t_i are defined as

$$a(\mu)\big|_{t_i} = \sum_{i}\sum_{j}\mu(a_{ij}), \quad i,j = 1,2,3,\ldots, M$$

Perimeter $P(\mu)$ is calculated as follows:

$$P(\mu) = \sum_{i=1}^{M}\sum_{j=1}^{M-1}\left|\mu(a_{ij}) - \mu(a_{i,j+1})\right| + \sum_{j=1}^{M}\sum_{i=1}^{M-1}\left|\mu(a_{ij}) - \mu(a_{i+1,j})\right|$$

6.3.3 Fuzzy Clustering Method

Jawahar and Ray [13] thresholded the image using the fuzzy c means clustering algorithm. Clustering is grouping similar objects into a group and dissimilar objects into another group. Thresholding may be looked upon as clustering where the image may be clustered into two or three or more depending on the grey level of objects. Fuzzy clustering may be looked as partitioning a set of *'n'* sample points $X = \{x_1, x_2, x_3, ..., x_n\}$ into *'c'* classes. If h_i is the histogram and p_i is the probability of the distribution of the grey values, $i \in \{0, 1, 2, 3, ..., L - 1\}$, the cluster means for each class is

$$v_j = \frac{\sum_{i=0}^{L-1} h_i i \mu_j(i)^\tau}{\sum_{i=0}^{L-1} h_i \mu_j(i)^\tau} \tag{6.3}$$

$j = 1, 2$ for the background and object region. The objective function $J = \sum_{j=1}^{2} \sum_{i=0}^{L-1} h_i \mu_j(i)^\tau d(i, v_j)^2 - \tau \geq 1$ controls the fuzziness in partition – can be iteratively minimized by computing the means from Equation 6.3 and updating the membership as

$$\mu_O(i) = \frac{1}{1 + \left[d(i, v_O)/d(i, v_B) \right]^{2/(\tau-1)}}$$

and

$$\mu_B(i) = 1 - \mu_O(i)$$

where *d* represents any distance function between the grey level *i* and class mean.

The mean and membership values are updated iteratively until there is no appreciable change in μ_O and μ_B.

As said earlier that as medical images contain uncertainties, fuzzy methods are very useful. So, examples on medical image thresholding using fuzzy divergence, fuzzy geometry and fuzzy clustering are shown.

Example 6.2

An example of thresholding medical images are shown in Figure 6.2 using fuzzy divergence, fuzzy geometry and fuzzy clustering on the medical images. Figure 6.2b is the result using fuzzy divergence, Figure 6.2c is the result using fuzzy geometry, and Figure 6.2d is the result using fuzzy clustering.

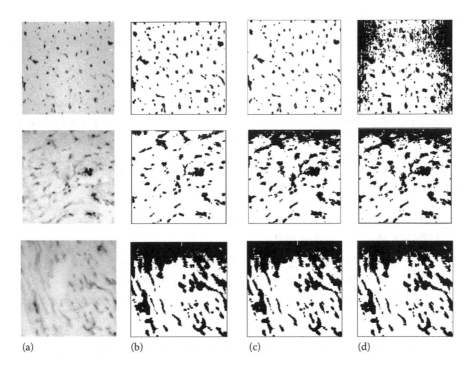

(a) (b) (c) (d)

FIGURE 6.2
(a) Blood vessel image, (b) fuzzy divergence, (c) fuzzy geometry and (d) fuzzy cluster.

It is observed from these figures that different fuzzy methods perform differently and work better on medical images, but in some cases as shown in Figure 6.2, blood vessels are not clearly visible due to noise present in the image. So, advanced fuzzy set theories are used by the researchers for thresholding where either more numbers of uncertainties or different kinds of uncertainties are used. This is done to obtain better results when better results are not obtained using fuzzy methods.

6.4 Intuitionistic Fuzzy Threshold Detection Methods

This section discusses different methods of image thresholding using intuitionistic fuzzy set theory. Fuzzy methods give better results as it deals with imprecise information (one uncertainty), but intuitionistic fuzzy set considers more (two) uncertainties. As medical images are poorly illuminated and the boundaries/edges and regions are not clearly visible, a better result may be expected using intuitionistic fuzzy set. There is very little work on thresh-olding of medical images, and few of these techniques are discussed in this section.

6.4.1 Intuitionistic Fuzzy Entropy–Based Method

Bustince et al. [2] suggested another thresholding scheme using intuitionistic fuzzy set theory. The membership function is defined using the restricted equivalent function.

A function $REF:[0, 1]^2 \rightarrow [0, 1]$ is called restricted equivalence function if it satisfies the following conditions [3]:

1. $REF(x, y) = REF(y, x)$ for all $x, y \in [0, 1]$.
2. $REF(x, y) = 1$, if and only if $x = y$.
3. $REF(x, y) = 0$, if and only if $x = 1, y = 0$ or $x = 0, y = 1$.
4. $REF(x, y) = REF(c(x), c(y))$, for all $x, y \in [0, 1]$, c being a strong negation.
5. For all $x, y, z \in [0, 1]$, if $x \leq y \leq z$, then $REF(x, y) \geq REF(x, z)$ and $REF(y, z) \geq REF(x, z)$.

They defined a restricted equivalence function as a measure of comparison:

$$REF(A, B) = 1 - |x - y| \tag{6.4}$$

Let c be a strong negation, such that $c(e) = e$ (e is an equilibrium point of negation) and let

$F: [0, 1] \rightarrow [e, 1]$ be a function such that

$F(x) = 1$ iff $x = 0$

$F(x) = e$ iff $x = 1$ and $F(x)$ is non-increasing

In this condition, the membership function at threshold 't' is defined as

$$\mu_B(t) = F\big(c\big(REF(g, m_B(t))\big)\big) \quad \text{if } g \leq t$$

$$\mu_O(t) = F\big(c\big(REF(g, m_O(t))\big)\big) \quad \text{if } g > t, \quad \text{'}g\text{' is the grey level} \tag{6.5}$$

where $m_O(t)$ and $m_B(t)$ are the average grey levels of the object and back-ground regions, respectively, and are given by the following formula:

$$m_O(t) = \frac{\sum_{g=0}^{t} g \cdot h(g)}{\sum_{g=0}^{t} h(g)}, \quad m_B(t) = \frac{\sum_{g=t+1}^{L-1} g \cdot h(g)}{\sum_{g=t+1}^{L-1} h(g)} \tag{6.6}$$

Let the function

$$F(x) = 1 - (1 - e)x \qquad (6.7)$$

with $F(0) = 1$, $F(1) = e$.

Substituting Equation 6.7 in Equation 6.5, the membership function is written as

$$\mu_B(t) = 1 - (1-e)c\big(REF(g, m_B(t))\big) \quad \text{if } g \leq t$$

$$\mu_O(t) = 1 - (1-e)c\big(REF(g, m_O(t))\big) \quad \text{if } g > t$$

If $c(x) = 1 - x$ for all $x \in [0, 1]$ and $REF(A, B) = 1 - |x - y|$ and $e = 0.5$, then

$$\mu_B(t) = 1 - 0.5|g - m_B(t)| \quad \text{if } g \leq t$$
$$\mu_O(t) = 1 - 0.5|g - m_O(t)| \quad \text{if } g > t \qquad (6.8)$$

Intuitionistic fuzzy set is constructed as

$$A_{IFS} = \left\{ \big(x, \mu_A(x)^\alpha, 1 - \mu_A^{1/\beta}(x)\big) \mid x \in X \right\}$$

The membership value $= \mu(g, T)^\alpha$ and the non-membership value $= 1 - \mu(g, T)^{1/\beta}$.

The optimal threshold is selected using Yager's intuitionistic fuzzy entropy:

$$IFE_A(t) = \frac{1}{M \times M} \sum_{g=0}^{L-1} h(g)\pi_A(g), \quad \pi_A(g) = 1 - \mu_A(g) - \nu_A(g)$$

So,

$$IFE_A(t) = \frac{1}{M \times M} \sum_{g=0}^{L-1} h(g)\big(\mu_t^{1/\beta}(g) - \mu_t^\alpha(g)\big) \qquad (6.9)$$

For each threshold grey level, t, intuitionistic fuzzy entropy is computed and the grey level corresponding to the minimum entropy is selected as the optimal threshold.

$\alpha = 4/3$ and $\beta = 5/4$ are used as the constant values.

With $\alpha > 6$ or $\beta > 6$, the results are not accurate, and with $\alpha = \beta = 1$, it becomes fuzzy.

If α is very large, the membership $\mu_A^\alpha(x) \approx 0$, which does not have any sense.

Couto et al. [10] suggested a thresholding method on general images. Consider an image A of size $M \times M$ with maximum grey level L and a_{ij} is the (i, j)th pixel of the image with $i, j = 1, 2, 3, \ldots, M$. Suppose $h(g)$ denotes the number of occurrences of the grey level g in the image. Given a certain threshold 't' that separates the objects and the background regions, the average grey levels of the object/background regions are computed using Equation 6.6.

The membership function is defined using a dissimilarity function that is expressed as

$$\mu_B(t) = F\left(d\left(\frac{g}{L-1}, \frac{m_B(t)}{L-1} \right) \right)$$

$$\mu_O(t) = F\left(d\left(\frac{g}{L-1}, \frac{m_O(t)}{L-1} \right) \right)$$

(6.10)

where from Equation 6.7, $F = 1 - 0.5x$, with $e = 0.5$, and from Equation 6.4, the restricted dissimilarity function is written as $d(x, y) = |x-y|$. Intuitionistic fuzzy set, π, is the ignorance of the expert in assigning the membership function, that is, the hesitation degree. The choice of the membership function depends on the expert's knowledge. If the expert is very certain about the pixels' belongingness to the object or background, then the hesitation degree, π, is zero and it decreases with the expert's certainty to the belongingness of the pixel. Also, the hesitation degree has the least possible influence on the choice of the membership degree. Under this condition, $\pi(g)$, the quantification of the ignorance of the expert in the selection of the membership function of the object and background is computed as

$$\pi(g) = (1 - \mu_B(g)) \cdot (1 - \mu_O(g))$$

An intuitionistic fuzzy entropy by Burillo and Bustince [1] is used as a measure to find the optimal threshold, which is defined as

$$IFE_B(t) = \frac{1}{M \times M} \sum_{g=0}^{L-1} h(g)\pi(g)$$

The grey level that corresponds to the lowest entropy is chosen as the optimal threshold. This can be expended to multilevel image segmentation.

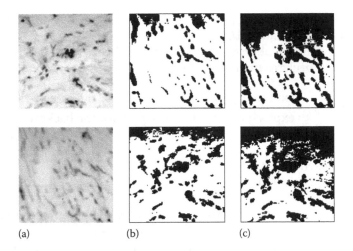

(a) (b) (c)

FIGURE 6.3
(a) Blood vessel image, (b) intuitionistic fuzzy method by Couto and (c) intuitionistic fuzzy method by Bustince.

Example 6.3

An example of a biomedical image is shown in Figure 6.3, which uses intuitionistic fuzzy entropy methods. Blood vessel images are very poorly illuminated, and thresholding becomes very difficult. Blood vessel thresholding is very important in pathology when there is a need to count the number of blood vessels in diagnosing diseases such as prostate cancer. Figure 6.3b is the intuitionistic fuzzy entropy method by Couto, and Figure 6.3c is the intuitionistic fuzzy entropy method by Bustince.

6.4.2 Intuitionistic Fuzzy Divergence–Based Method

Vlachos and Sergiadis [24] suggested an intuitionistic fuzzy image thresholding method that is similar to the idea as described by Chaira and Ray [5] where the minimization of actual and ideal thresholded image leads to maximum belongingness of foreground pixels to foreground regions and background pixels to background regions.

The gamma membership function is calculated from the gamma distribution [9,10], which is given as

$$f(x) = \exp(-(x - m)) \tag{6.11}$$

Let us consider an image A of size $M \times M$ with maximum grey level L and g_{ij} the (i, j)th pixel of the image with $i, j = 1, 2, 3, \ldots, M$. Then, given a certain threshold 't' that separates the objects and the background regions, the average grey levels of the object and background regions, $m_O(t)$ and $m_B(t)$, are computed using Equation 6.6.

Replacing m in Equation 6.11 by m_O and m_B, the membership function for the object and background regions becomes

$$\mu_A\left(g_{ij}\right) = \begin{matrix} \exp(-c \cdot |g_{ij} - m_O|), & \text{if } g_{ij} \le t, \text{for the object} \\ \exp(-c \cdot |g_{ij} - m_B|), & \text{if } g_{ij} > t, \text{for the background} \end{matrix} \quad (6.12)$$

where
t is any chosen threshold
g_{ij} is the (i, j)th pixel in image A

The constant 'c' is chosen to ensure that the membership of the grey level corresponds to the range [0, 1] and is calculated as $c = 1/(g_{max} - g_{min})$, where g_{min} and g_{max} are the minimum and maximum values of the grey level in the image, respectively.

Based on the fuzzy set, the membership and non-membership functions in an intuitionistic fuzzy domain are constructed as follows:

$$\text{Membership value,} \quad \mu_{IFS}(g, T) = \lambda \mu(g, T)$$

$$\text{Non-membership value,} \quad v_{IFS}(g, T) = (1 - \mu(g, T))^{\lambda}$$

where
g is the grey level of the image
T is a certain threshold
$\lambda \in [0, 1]$, and in this method, $\lambda = 0.2$ is selected

From the intuitionistic fuzzy discrimination measure derived from the cross entropy,

$$D_{IFS}(A, B) = I_{IFS}(A, B) + I_{IFS}(B, A)$$

$$I_{IFS}(A, B) = \sum_i \left(\mu_A(x_i) \cdot \ln \frac{\mu_A(x_i)}{(1/2)(\mu_A(x_i) + \mu_B(x_i))} + v_A(x_i) \ln \frac{v_A(x_i)}{(1/2)(v_A(x_i) + v_B(x_i))} \right)$$

In the image, the intuitionistic fuzzy discrimination information is

$$D_{IFS}(A, B) = \sum h(g) \left(\mu_A(g) \cdot \ln \frac{2\mu_A(g)}{\mu_A(g) + \mu_B(g)} + v_A(g) \cdot \ln \frac{2v_A(g)}{v_A(g) + v_B(g)} \right)$$
$$+ \sum h(g) \left(\mu_B(g) \cdot \ln \frac{2\mu_B(g)}{\mu_B(g) + \mu_A(g)} + v_B(g) \cdot \ln \frac{2v_B(g)}{v_A(g) + v_B(g)} \right) \quad (6.13)$$

The image is compared with an ideally thresholded image. Considering B to be an ideally thresholded image where all the object/background pixels perfectly belong to their respective regions, the membership function of object pixels, $\mu_B(g) = 1$, Equation 6.13 is rewritten as

$$D_{IFS}(A,B) = \sum_g h(g)\left(\mu_A(g)\cdot\ln\frac{2\mu_A(g)}{1+\mu_A(g)}+\nu_A(g)\ln 2+\ln\frac{2}{1+\mu_A(g)}\right) \quad (6.14)$$

Divergence is calculated for all the threshold grey levels. The optimal threshold is the grey value corresponding to the minimum divergence value.

Chaira [7] suggested a divergence-based method for medical image thresholding. The membership function is obtained using Equation 6.10 as

$$\mu_A\left(a_{ij}\right) = \begin{matrix} \exp(-c\cdot|g_{ij}-m_O|), & \text{if } g_{ij} \le t, \text{for the object} \\ \exp(-c\cdot|g_{ij}-m_B|), & \text{if } g_{ij} > t, \text{for the background} \end{matrix}$$

where
$m_O(t)$ and $m_B(t)$ are computed using Equation 6.6

The constant 'c' is $c = 1/(g_{max} - g_{min})$ as described earlier.

The non-membership function is computed using Sugeno-type intuitionistic fuzzy generator [21].

Sugeno's intuitionistic fuzzy generator is written as

$$N(\mu(x)) = \frac{(1-\mu(x))}{(1+\lambda\mu(x))}, \quad \lambda > 0 \quad (6.15)$$

with hesitation degree

$$\pi_A(x) = 1 - \mu_A(x) - \frac{(1-\mu_A(x))}{(1+\lambda\cdot\mu_A(x))} \quad (6.16)$$

Intuitionistic fuzzy divergence suggested by Chaira and Ray [4] in section 3 is used for finding the optimal threshold. A brief idea on the intuitionistic fuzzy divergence measure is detailed below.

6.4.2.1 Intuitionistic Fuzzy Divergence Measure

The intuitionistic fuzzy divergence measure considers three parameters, namely, the membership degree, the non-membership degree and the hesitation degree (or intuitionistic fuzzy index).

Let $A = \{(x, \mu_A(a_{ij}), \nu_A(a_{ij})) | a_{ij} \in A\}$ and $B = \{(x, \mu_B(b_{ij}), \nu_B(b_{ij})) | b_{ij} \in B\}$ be two intuitionistic fuzzy images. If $\mu_A(a_{ij})$ and $\nu_A(b_{ij})$ are the membership and non-membership values of (i, j)th element of image A, respectively, then while assigning the membership value of an element, there may be some hesitation degree $\pi_A(a_{ij})$, where $\pi_A(a_{ij}) = 1 - \mu_A(a_{ij}) - \nu_A(a_{ij})$. In that case, the membership value will lie in the range $\{\mu_A(a_{ij}), \mu_A(a_{ij}) + \pi_A(a_{ij})\}$. Similarly for image B, the membership value will also lie in a range.

So, the divergence between pixels a_{ij} and b_{ij} of images A and B is given as

$$D_{IFS}(A,B)$$

$$= \sum_i \sum_j \Big[2 - \big(1 - \mu_A(a_{ij}) + \mu_B(b_{ij})\big) \cdot e^{\mu_A(a_{ij}) - \mu_B(b_{ij})}$$

$$- \big(1 - \mu_B(b_{ij}) + \mu_A(a_{ij})\big) \cdot e^{\mu_B(b_{ij}) - \mu_A(a_{ij})}$$

$$+ 2 - \big(1 - \mu_A(a_{ij}) - \pi_A(a_{ij}) + \mu_B(b_{ij}) + \pi_B(b_{ij})\big) \cdot e^{\mu_A(a_{ij}) + \pi_A(a_{ij}) - \mu_B(b_{ij}) - \pi_B(b_{ij})}$$

$$- \big(1 - \mu_B(b_{ij}) - \pi_B(b_{ij}) + \mu_A(a_{ij}) + \pi_A(a_{ij})\big) \cdot e^{\mu_B(b_{ij}) + \pi_B(b_{ij}) - \mu_A(a_{ij}) - \pi_A(a_{ij})} \Big]$$

$$(6.17)$$

For each threshold, the thresholded image is compared with an ideally thresholded image where the background and foreground regions are precisely segmented. In an ideally segmented image, the membership, non-membership and hesitation degrees of all pixels are 1, 0 and 0, respectively. Thus, with $\mu_B(b_{ij}) = 1$ and $\pi_B(b_{ij}) = 0$, the IFD in Equation 6.17 is reduced to

$$D_{IFS}(A,B) = \sum_i \sum_j \Big[2 - (1 - \mu_A(a_{ij}) + 1) \cdot e^{\mu_A(a_{ij}) - 1} - (1 - 1 + \mu_A(a_{ij})) \cdot e^{1 - \mu_A(a_{ij})}$$

$$+ 2 - (1 - \mu_A(a_{ij}) - \pi_A(a_{ij}) + 1 + 0) \cdot e^{\mu_A(a_{ij}) + \pi_A(a_{ij}) - 1 - 0}$$

$$- (1 - 1 - 0 + \mu_A(a_{ij}) + \pi_A(a_{ij})) \cdot e^{1 + 0 - \mu_A(a_{ij}) - \pi_A(a_{ij})} \Big]$$

$$\Leftrightarrow \qquad\qquad\qquad\qquad\qquad\qquad\qquad\qquad\qquad\qquad\qquad (6.18)$$

$$D_{IFS}(A,B) = \sum_i \sum_j \Big[2 - (2 - \mu_A(a_{ij})) \cdot e^{\mu_A(a_{ij}) - 1} - \mu_A(a_{ij}) \cdot e^{1 - \mu_A(a_{ij})}$$

$$+ 2 - (2 - \mu_A(a_{ij}) - \pi_A(a_{ij})) \cdot e^{\mu_A(a_{ij}) + \pi_A(a_{ij}) - 1}$$

$$- (\mu_A(a_{ij}) + \pi_A(a_{ij})) \cdot e^{1 - \mu_A(a_{ij}) - \pi_A(a_{ij})} \Big]$$

The value of π is calculated from $\pi_A(a_{ij}) = 1 - \mu_A(a_{ij}) - \nu_A(a_{ij})$. $\nu_A(a_{ij})$ is computed using Equation 6.15.

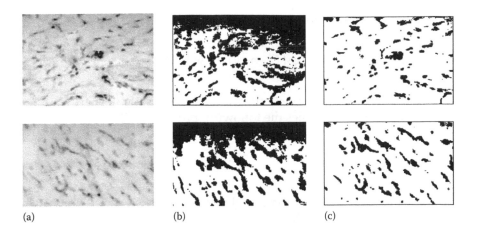

(a)   (b)   (c)

FIGURE 6.4
(a) Blood vessel image, (b) intuitionistic fuzzy method by Vlachos and (c) intuitionistic fuzzy method by Chaira.

For all threshold grey levels, the intuitionistic fuzzy divergence is calculated. The threshold level corresponding to the minimum divergence is selected as the optimal threshold.

Example 6.4

Two examples of medical images (by Vlachos and Chaira) are shown in Figure 6.4 to illustrate the efficacy of intuitionistic fuzzy divergence thresholding methods. Figure 6.4b is the result using Vlachos' method, and Figure 6.4c is the result using Chaira's method.

6.5 Window-Based Thresholding

Local thresholding is useful in medical images as it selects the threshold for each window. As medical images are poorly illuminated, global threshold may not work better. In pathological images, where the blood vessels are hardly visible, window-based thresholding will work better. In window-based thresholding, the threshold is selected for each window, so the image will be thresholded depending on the image characteristics for that window.

Chaira [8] suggested a window-based method for image thresholding. In this method, the restricted equivalence function is used to find the membership function.

If φ_1 and φ_2 are two automorphisms in a unit interval, then

$$REF(x,y) = \varphi_1^{-1}\left(1 - |\varphi_2(x) - \varphi_2(y)|\right) \quad \text{with } c(x) = \varphi_2^{-1}\left(1 - \varphi_2(x)\right)$$

is a restricted equivalence function. '*c*' is a strong negation, where $c: [0, 1] \rightarrow [0, 1]$ is a negation if it satisfies the following properties [3]:

1. $c(0) = 1, c(1) = 0$
2. $c(x) \leq c(y)$, iff $x \geq y$

and automorphism in an interval $[a, b]$ is a continuously strictly increasing function $\varphi: [a, b] \rightarrow [a, b]$ such that $\varphi(a) = a, \varphi(b) = b$.

6.5.1 Calculation of Membership Function

Let us consider $\varphi_2(x) = x$. From the definition of restricted equivalence function,

$$REF(x,y) = \varphi_1^{-1}(1 - |\varphi_2(x) - \varphi_2(y)|) = \varphi_1^{-1}(1 - |x - y|)$$

So, $REF(x,y) = \varphi_1^{-1}(1 - |x - y|)$.

Considering $\varphi_1(x) = \ln[x(e - 1) + 1]$, with $e = \exp(1)$ and using inverse function, we get

$$\varphi_1^{-1}(x) = \frac{(e^x - 1)}{(e - 1)}$$

Thus,

$$REF(x,y) = \frac{(e^{1-|x-y|} - 1)}{(e - 1)} \tag{6.19}$$

This equation is used to define the membership function.

The membership function denotes the belongingness of a pixel to a region. So, the smaller the difference between the grey level of any pixel a_{ij} and the mean of the region to which the pixel belongs, the greater the membership value and vice versa.

Let us define the membership function $\mu: [0, 1]$ as

$$\mu(x) = REF(x, y)$$

Then using Equation 6.19, the membership function becomes

$$\mu(x) = 0.582 \cdot (e^{1-|x-y|} - 1) \tag{6.20}$$

The value 0.582 is computed from $1/(\exp(1) - 1) = 0.582$.

For a certain threshold 't' that separates the object and background regions, the membership function of the object region is written using Equation 6.20 as

$$\mu_A(a_{ij}) = 0.582(e^{1-|a_{ij}-m_O|} - 1), \quad \text{if } a_{ij} \le t, \text{ for the object}$$

Likewise, the membership function $\mu_A(a_{ij})$ of the background region is written as

$$\mu_A(a_{ij}) = 0.582(e^{1-|a_{ij}-m_B|} - 1), \quad \text{if } a_{ij} > t, \text{ for the background} \tag{6.21}$$

with $m_O(t)$ and $m_B(t)$ are computed using Equation 6.6, respectively.

The Sugeno-type intuitionistic fuzzy generator is used to find the non-membership function in an intuitionistic fuzzy domain that is written as

$$v(x) = \frac{1-\mu_A(x)}{1+\lambda \cdot \mu_A(x)}, \quad \lambda > 0$$

For thresholding the images, the image is initially filtered with a Gaussian filter of size 3×3. The filtered image is divided into several windows $(1/4)^*$ the image size. With smaller window size, the threshold picks very small particles, which results in poor extraction and performance of the image.

The membership values are calculated for each window, and the non-membership values are also computed using Sugeno-type intuitionistic fuzzy generator. The hesitation degree is calculated using the membership and non-membership values. The value of λ in the construction of an intuitionistic fuzzy set is taken as $\lambda = 0.8$. For each threshold, fuzzy divergence between an ideal and actual thresholded images is calculated. The divergence measure between the ideal and actual thresholded images is computed using Equation 6.18 as

$$D_{IFS}(A,B) = \sum_i \sum_j \left[2 - (2 - \mu_A(a_{ij})) \cdot e^{\mu_A(a_{ij})-1} - \mu_A(a_{ij}) \cdot e^{1-\mu_A(a_{ij})} \right.$$

$$+ 2 - (2 - \mu_A(a_{ij}) - \pi_A(a_{ij})) \cdot e^{\mu_A(a_{ij})+\pi_A(a_{ij})-1}$$

$$\left. - (\mu_A(a_{ij}) + \pi_A(a_{ij})) \cdot e^{1-\mu_A(a_{ij})-\pi_A(a_{ij})} \right] \tag{6.22}$$

For all threshold grey levels, intuitionistic fuzzy divergence is calculated. The threshold level corresponding to the minimum divergence is selected as the optimal threshold.

This is calculated for all the windows. The final threshold is the optimal threshold minus (1/4)th of the standard deviation of the window:

$$\text{Final\_Th} = \text{Th}^{opt}_{window} - \left(\frac{1}{4}\right) \times \text{S.D.}_{window}$$

where S.D. is the standard deviation

Had the fraction not been there, the images will be more segmented, that is, all the blood vessels are not extracted. Thus, the fraction of 1/4 has been used to obtain a good segmented image.

Example 6.5

Two examples of blood vessel and blood cell images are shown in Figure 6.5 to illustrate the efficacy of window-based thresholding using intuitionistic fuzzy set on medical images. It is observed that the blood vessels/cells are clearly thresholded on poor contrasted medical images.

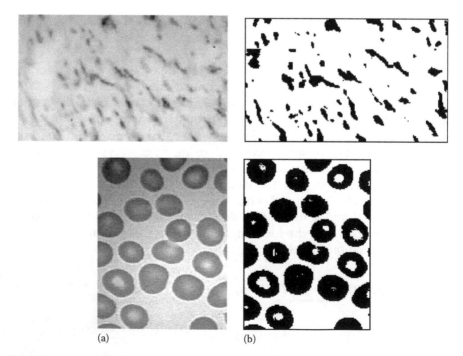

(a)   (b)

FIGURE 6.5
(a) Blood vessel image and (b) intuitionistic fuzzy local thresholding method by Chaira.

6.6 Thresholding Using Type II Fuzzy Set Theory

If we interpret images in terms of Type II fuzzy set, then a question will arise as how fuzziness is a fuzzy set, that is, to what degree the membership values are certain. If the degrees are defined with no uncertainty, then this fuzziness diminishes. Tizhoosh [22] defined this fuzziness in terms of ultrafuzziness where maximum ultrafuzziness is 1. But if the membership value lies in an interval range, then this ultrafuzziness will increase. Ultrafuzziness is defined as

$$\gamma(A) = \frac{1}{M \times N} \sum_{g=0}^{L-1} h(g)[\mu_U(g) - \mu_L(g)] \qquad (6.23)$$

where
   $\mu_U(g)$ and $\mu_L(g)$ are the upper and lower membership values of the interval, respectively
   'g' is the grey level of the image

This ultrafuzziness is used in thresholding. The following cases hold for the ultrafuzziness:

1. If μ_A is a type 1 or ordinary fuzzy set, then $\mu_U(g) = \mu_L(g)$ and then $\gamma(A) = 0$.
2. If $\mu_U(g) - \mu_L(g) = 1$, then $\gamma(A) = 1$.
3. $\gamma(A) \geq \gamma(A)'$ if A' is a crisper version of A.

Tizhoosh [22] suggested Type II thresholding using the measure of ultrafuzziness. Type II fuzzy set considers the membership function as fuzzy and so the membership function lies in an interval. As the membership function is considered to be fuzzy, better results may be expected on medical images. A measure of ultrafuzziness is used to find the optimal threshold. The concept of ultrafuzziness aims at capturing/eliminating the uncertainties within fuzzy systems using type I fuzzy sets. A Type II fuzzy set may be written as

$$A_{\text{TypeII}} = \{x, \mu_U(x), \mu_L(x) \mid x \in X\}$$

and

$$\mu_U(x) < \mu(x) < \mu_L(x), \quad \mu \in (0,1]$$

From Equation 6.23, the upper and lower limits are written as

$$\mu_U(x) = \mu(x)^{1/\alpha}$$

$$\mu_L(x) = \mu(x)^\alpha$$

(6.24)

respectively, with $\alpha \in [1, 2]$.

Initially, the image is fuzzified with any user-defined membership function and α is selected. Then the lower and upper values of the membership function are computed using Equation 6.24. For each threshold grey level, the measure of ultrafuzziness is computed as well as the maximum value. The threshold grey level that agrees with the maximum ultrafuzziness value corresponds to the optimal threshold of the image.

Example 6.6

Two examples of blood vessel and blood cell images are shown to illustrate the efficacy of the Type II fuzzy method (Figure 6.6).

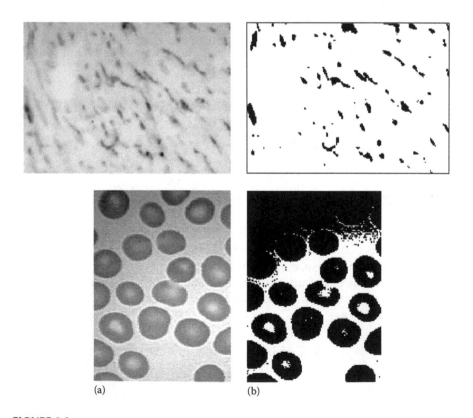

(a)   (b)

FIGURE 6.6
(a) Blood vessel image and (b) Tizhoosh Type II fuzzy method.

6.7 Segmentation Using Type II Fuzzy Set Theory

Tizhoosh also suggested a segmentation method [23] using Type II fuzzy set, but on artificial images. In most of the images, the object of interest is dark and the primary membership of dark pixels is defined using a Z membership function in an interval $(g_{max}-((g_{max} + g_{min})/2))$. So,

$$\mu_{A(g)} = \begin{cases} 1-8\left(\dfrac{g-g_{min}}{g_{max}-g_{min}}\right)^2 & \text{if } g_{min} \le g \le \dfrac{g_{max}+g_{min}}{4} \\[3mm] 2\left(\dfrac{\left(g_{max}-g_{min}\right)\left(g_{max}+g_{min}\right)-8g}{g_{max}-g_{min}}\right)^2 & \text{if } \dfrac{g_{max}+g_{min}}{4} < g < \dfrac{g_{max}+g_{min}}{2} \\[3mm] 0 & \text{if } g > \dfrac{g_{max}+g_{min}}{2} \end{cases}$$

$$(6.25)$$

The lower and upper membership functions are computed as

$$\mu_L(i,j) = \mu\left(g(i,j)\right)^{\alpha(i,j)}$$

$$\mu_U(i,j) = \mu\left(g(i,j)\right)^{1/\alpha(i,j)}$$

α for each point is computed with respect to its neighbourhood:

$$\alpha(i,j) = F \times \min\left(1, \frac{\max_k g(i+k,j+k)-\min_k g(i+k,j+k)}{L-1}\right) \quad (6.26)$$

where
$k \in [-n, ..., -1, 0, 1, ..., n]$, $n = W/2$, and W is the size of the window
F is an amplification factor that lies between $[1, \infty]$

With more F value, that is, around 20, weak edges also are included, and with lesser F value, that is, around 2, strong edges are included.

This means that if the difference in the intensity of the centre and its immediate neighbourhood is more, then uncertainty increases. Next, the weight of upper and lower membership functions is computed as

$$w_L(i,j) = \frac{\min_k g(i+k,j+k)}{g_{max}}$$

$$w_U(i,j) = \frac{\max_k g(i+k,j+k)}{g_{max}}$$

Finally, the segmented pixel g^{seg} is computed as follows:

$$g^{seg}(i,j) = (L-1) \times \frac{w_L(i,j)\mu_L(g(i,j)) + w_U(i,j)\mu_U(g(i,j))}{w_L(i,j) + w_U(i,j)} \qquad (6.27)$$

6.8 Segmenting Leucocyte Images in Blood Cells

Leucocyte segmentation is a special type of segmentation where different types of leucocytes are segmented with shapes preserved. Thresholding leucocytes is a difficult task. In pathological studies, blood cell parameters such as red blood cells, white blood cells and platelets, erythrocytes and leucocytes are very essential to detect many diseases such as anaemia, leukaemia, cancer and any other infection. Out of these blood parameters, leucocyte (white blood cell) counting is very much essential where counting five types of leucocytes such as eosinophil, basophil, neutrophil, lymphocytes and monocytes is done to detect diseases. The five types of leucocytes can be distinguished by their cytoplasmic granules, the staining properties of the granules, the size of cell, the proportion of the nucleus to the cytoplasmic material and the type of nucleolar lobes.

To threshold the leucocytes, any thresholding method can be used, but the method should be such that it should preserve the shape of the leucocytes. The shape of the nucleus is preserved to distinguish different types of leucocytes to diagnose diseases. For this, interval Type II fuzzy set and a modified Cauchy distribution are used to find the membership values of the image. The optimal threshold is selected by using divergence using Type II fuzzy set [9].

6.8.1 Cauchy Distribution

The probability density function of Cauchy distribution is

$$f(x;a,\gamma) = \frac{1}{\pi\gamma\left(1 + \left(\dfrac{x-a}{\gamma}\right)^2\right)}$$

where
  a is the location parameter
  γ is a scale parameter

For standard Cauchy distribution, $\gamma = 1$ and $a = 0$, and the distribution becomes

$$f(x,0,1) = \frac{1}{\pi(1+x^2)}$$

To find the membership values of the image, the distribution is modified. Considering

$$\gamma = \frac{1}{\sqrt{const}}$$

and

const $= 1/(f_{max} - f_{min})$, where f_{max} and f_{min} are the maximum and minimum grey values of the image, respectively, and substituting the values, the distribution becomes

$$f(x;a,\gamma) = \frac{\sqrt{const}}{\pi\left(1+const\cdot(x-a)^2\right)} = \left(\frac{\sqrt{const}}{\pi}\right)\frac{1}{\left(1+const\cdot(x-a)^2\right)}$$

In order to make the membership values feasible, the constant term $\left(\sqrt{const}/\pi\right)$ is not taken into account. So, the function that is used to derive the membership function is

$$f(x,a) = \frac{1}{\left(1+const\cdot(x-a)^2\right)} \tag{6.28}$$

In segmentation, *t*-conorm is used to form a new membership function using the two membership functions – upper and lower membership levels in Type II fuzzy set.

The image is initially fuzzified using the modified Cauchy membership function in Equation 6.28.

Then the upper and lower membership functions of interval Type II fuzzy set for the object and background regions are computed as

$$\mu^{lower}(g_{ij}) = \left[\mu(g_{ij})\right]^{1/\alpha}$$

$$\mu^{upper}(g_{ij}) = \left[\mu(g_{ij})\right]^{\alpha}$$

where

$$\mu(g) = \frac{1}{\left(1+const\cdot(x-m_O)^2\right)}, \quad object$$

$$= \frac{1}{\left(1+const\cdot(x-m_B)^2\right)}, \quad background$$

m_O and m_B are computed using Equation 6.6, respectively.

Using the probabilistic *t*-conorm, a new membership function is generated, which is written as

$$\mu_P(g_{ij}) = \mu^{\text{upper}}(g_{ij}) + \mu^{\text{lower}}(g_{ij}) - \mu^{\text{upper}}(g_{ij}) \cdot \mu^{\text{lower}}(g_{ij}) \tag{6.29}$$

For each threshold grey level, fuzzy divergence is computed between the thresholded image and ideally segmented image. Ideally segmented image is such that the membership values are all 1 [5,6]:

$$D(A,B) = \sum_{j=0}^{M-1} \sum_{i=0}^{N-1} \Big[2 - (1 - \mu_A(g_{ij}) + \mu_B(g_{ij})) \cdot e^{\mu_A(g_{ij}) - \mu_B(g_{ij})}$$

$$- (1 - \mu_B(g_{ij}) + \mu_A(g_{ij})) \cdot e^{\mu_B(g_{ij}) - \mu_A(g_{ij})} \Big] \tag{6.30}$$

Substituting $\mu_B(g_{ij}) = 1$, for an ideally thresholded image, we get

$$D(A,B) = \sum_{j=0}^{M-1} \sum_{i=0}^{N-1} \Big[2 - (2 - \mu_A(g_{ij})) \cdot e^{\mu_A(g_{ij}) - 1} - \mu_A(g_{ij}) \cdot e^{1 - \mu_A(g_{ij})} \Big]$$

where
$$\mu_A(g_{ij}) = \mu_P^A(g_{ij})$$
$$\mu_B(g_{ij}) = \mu_P^B(g_{ij})$$

Fuzzy divergence is computed for all the threshold grey levels. The threshold grey level at which the fuzzy divergence is minimum is the optimal threshold. For better segmentation, the threshold is chosen as half of the optimal threshold.

Example 6.7

Two examples of leucocyte images are shown in Figures 6.7 and 6.8. In one experiment, images contain a normal cell, and in another experiment, leukemic cell images are used that contain mature and immature cells. In leukaemia, there is an increase in the number of immature/abnormal white blood cells. Segmentation becomes difficult when many leucocytes (mature or immature) are present in the image and cells and are jumbled together. It is observed that the shapes of the leucocytes are properly preserved and are accurately distinguishable.

Performance evaluation: In order to verify the performance of segmentation methods, a ground truth image or manually segmented images are drawn.

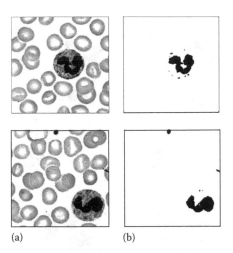

(a)   (b)

FIGURE 6.7
(a) Normal leucocyte images and (b) segmented leucocyte images using Type II fuzzy set. (From *Micron*, 61, Chaira, T., Accurate segmentation of leukocyte in blood cell images using Atanassov's intuitionistic fuzzy and interval Type II fuzzy set theory, 1–4, Copyright 2014, with permission from Elsevier.)

Obtaining a ground truth image is very tedious and not perfect, but still it gives a useful indication. For all the methods, the misclassification error is calculated, which is defined as [25]

$$\text{Error} = 1 - \frac{\left|B_{ET} \cap B_{GT}\right| + \left|F_{ET} \cap F_{GT}\right|}{B_{GT} + F_{GT}}$$

where
  F_{GT} and B_{GT} denote the foreground and background area pixels of the ground truth image, respectively
  F_{ET} and B_{ET} are the foreground and background area pixels of the experimental thresholded image, respectively

This error reflects the percentage of wrongly assigned pixels that ranges from zero or no error, that is, when the image is exactly segmented to 1 or when the image is wrongly segmented. The total area of the object pixels, which are the cells shown in black, is obtained from the histogram of the segmented image, and likewise, the background (white) pixels are computed from the histogram of the segmented image. The unit of area is in pixels. If the area covered by each pixel is known, that is, x (mm²/pixel), then it is possible to convert to a physical unit. Ideally, the misclassification error is zero.

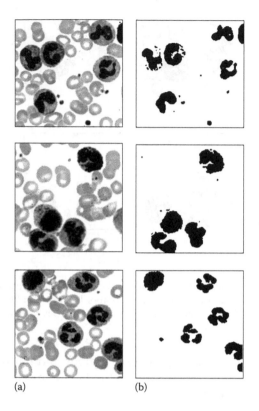

(a)   (b)

FIGURE 6.8
(a) Abnormal leucocyte images and (b) segmentation leucocyte images using Type II fuzzy set. (From *Micron*, 61, Chaira, T., Accurate segmentation of leukocyte in blood cell images using Atanassov's intuitionistic fuzzy and interval Type II fuzzy set theory, 1–4, Copyright 2014, with permission from Elsevier.)

6.9 Examples Using MATLAB®

A MATLAB® code for image thresholding is given, which will be beneficial to the readers to implement the method.

6.9.1 Intuitionistic Windowed Thresholding Method

```
a1=imread('bloodcell.jpg');
a2=rgb2gray(a1);
dim1=120 ;dim2=160;          % size of the image
im1=imcrop(a2,[2 1 dim2-1 dim1-1]);
```

```
  dim3=dim1/8;dim4=dim2/8;
  %computing mean for each window based
mun2=[];
mun=[];
  for i1=1:dim3:dim1
    mun1=[];
    for i2=1:dim4:dim2
  img1=imcrop(im1,[i2 i1 dim4-1 dim3-1]);
img=double(img1);
    mx=max(max(img))+1;
    mn=min(min(img))+1;
    c=double(fix(img));size(c);
stdf1=std2(img);
for p=mn:mx
   s=0.0;
   for b=mn:p
   z=count(b)+ s;
   s=z;
   z;
end
z;
s1=0.0;
for b=p+1:mx
   if p<mx
       z1=count(b)+s1;
       s1=z1;
       z1;
     else
  z1=0.0;
   end
z1;
end
z1;
t=0.0;
for b=mn:p
    m =count(b).*(x(b)-1) +t;
   t = m;
 end
m;
m1=fix(m*(1/z));   % mean of the object region
u=0.0;
for b=p+1:mx
  if p<mx
      n = count(b).*(x(b)-1)+ u;
      u = n;
else
   n=0.0;
end
end
n;
```

```
con1=0.8;
n1=fix((1/z1)*n);    % mean of the background region
%computing the membership values of pixels in each window
   using restricted equivalence function
for i=1:dim3
  for j=1:dim4
     if c(i,j)<p
       mem(i,j) = 0.582*(exp(1-abs(c(i,j)-m1)/255)-1);
       mem(i,j)= mem(i,j);
       nonmem(i,j)= (1-mem(i,j))/(1+con1*mem(i,j));
       hes(i,j)=1-mem(i,j)-nonmem(i,j);
       mem2(i,j)= mem(i,j)+ hes(i,j);
         else
       mem(i,j) = 0.582*(exp(1-abs(c(i,j)-n1)/255)-1);
       mem(i,j)= mem(i,j);
       nonmem(i,j)= (1-mem(i,j))/(1+con1*mem(i,j));
         hes(i,j)=1-mem(i,j)-nonmem(i,j);
         mem2(i,j)= mem(i,j)+ hes(i,j);
                end
     end
  end
% finding optimum threshold using intuitionistic fuzzy divergence
  u1=0.0;
for i=1:dim3
  for j=1:dim4
     h= 2-(2-mem(i,j))*exp(mem(i,j)-1)-mem(i,j)*exp(1-mem(i,j))+...
       2-(2-mem2(i,j))*exp(mem2(i,j)-1)-(mem2(i,j))*exp(1-
         mem2(i,j))+u1;
       u1=h;
     end
  end
  div=h;
  div_vec = [mun;div];
  mun = div_vec;
  h1(p)= div;
end
div_vec;
  l=min(h1(mn:mx));
  [p]=find(l==h1);
  medp=mean(p);  % optimum threshold
 imgbw=zeros(size(img));
imgbw(find(img>medp-0.25*stdf1))=1;
finim=[mun1,imgbw];
mun1=finim;
      end
 finimage1=[mun2;finim];
 mun2=finimage1;
end
 figure,imshow(finimage1)
```

6.10 Summary

In this chapter, various image thresholding schemes using advanced fuzzy sets such as intuitionistic fuzzy and Type II fuzzy set theories to segment poor contrasted blood vessels and blood cells are discussed. Fuzzy set theory does give better results, but in some cases when fuzzy methods fail to provide better results, advanced fuzzy set theoretic techniques may be useful as these sets consider more uncertainties. A special type of segmentation, that is, leucocyte segmentation, is also discussed where the leucocytes are thresholded with shapes preserved. It is very much essential in diagnosing different kinds of diseases. Also, MATLAB coding is provided, which will be helpful to the readers in implementing the algorithm.

References

1. Burillo, P. and Bustince, H., Entropy on intuitionistic fuzzy sets and on interval-valued, *Fuzzy Sets and Systems*, 78, 305–316, 1996.
2. Bustince, H., Mohedano, V., Barrenechea, E., and Pagola, M., An algorithm for calculating the threshold of an image representing the uncertainty through A-IFS, in *Proc. of the 11th IPMU*, 2383–2390, Paris, France, 2006.
3. Bustince, H., Barrenechea, E., and Pagola, M., Restricted equivalence function, *Fuzzy Sets and Systems*, 157(17), 2333–2346, 2006.
4. Chaira, T. and Ray, A.K., A new measure on intuitionistic fuzzy set and its application to edge detection, *Applied Soft Computing*, 8(2), 919–927, 2008.
5. Chaira, T. and Ray, A.K., Segmentation using fuzzy divergence, *Pattern Recognition Letters*, 24(12), 1837–1844, 2006.
6. Chaira, T. and Ray, A.K., Thresholding using fuzzy set theory, *Pattern Recognition Letters*, 24(12), 1837–1844, 2003.
7. Chaira, T., Medical image thresholding scheme using intuitionistic fuzzy set, in *Proc. of IEEE, ICM2CS'09*, 1–5, JNU, New Delhi, India, 2009.
8. Chaira, T., Intuitionistic fuzzy segmentation of medical images, *IEEE Transactions of Biomedical Engineering*, 57(6), 1430–1436, 2010.
9. Chaira, T., Accurate segmentation of leukocyte in blood cell images using Atanassov's intuitionistic fuzzy and interval Type II fuzzy set theory, *Micron*, 61, 1–4, 2014.
10. Couto, P. et al., Image segmentation using A-IFS, in *Processing of IPMU*, Malaga, Spain, pp. 1620–1627, 2008.
11. Couto, P. et al., Uncertainty in multilevel thresholding using Atanassov's intuitionistic fuzzy sets, in *IEEE World Congress on Computational Intelligence*, Hong Kong, China, pp. 330–335, 2008.
12. Huang, L.K. and Wang, M.J., Image thresholding by minimizing the measure of fuzziness, *Pattern Recognition*, 28(1), 41–51, 1995.

13. Jawahar, C.V. and Ray, A.K., Investigations on fuzzy thresholding based on fuzzy clustering, *Pattern Recognition*, 10(10), 1605–1613, 1997.
14. Niblack, W., *An Introduction to Image Processing*, Prentice Hall, Englewood Cliffs, NJ, 1986.
15. Otsu, N., A threshold selection method from gray level histograms, *IEEE Transactions on Systems, Man, and Cybernetics*, 9, 62–66, 1979.
16. Pal, S.K. and Rosenfeld, A., Image enhancement and thresholding by optimization of fuzzy compactness, *Pattern Recognition Letters*, 7, 77–86, 1998.
17. Rosenfeld, A., The fuzzy geometry of image subsets, *Pattern Recognition Letters*, 2, 311–317, 1984.
18. Rodriguez, R., A strategy for blood vessel segmentation based on threshold which combines statistical and scale space filter: Application to study of angiogenesis, *Computer Methods and Programs in Biomedicine*, 80, 1–9, 2006.
19. Sankur, B. and Sezgin, M., Survey over image thresholding techniques and quantitative performance evaluation, *Journal of Electronic Imaging*, 13(1), 146–165, 2004.
20. Sahoo, P.K. et al., A survey of thresholding techniques, *Computer Vision, Graphics and Image Processing*, 41(2), 233–260, 1988.
21. Sugeno, M., Fuzzy measures and fuzzy integral: A survey, in *Fuzzy Automata and Decision Processes*, M.M. Gupta, G.S. Sergiadis, and B.R. Gaines (Eds.), North Holland, Amsterdam, the Netherlands, 89–102, 1977.
22. Tizhoosh, H.R., Image thresholding using type II fuzzy sets, *Pattern Recognition*, 38, 2363–2372, 2005.
23. Tizhoosh, H.R., Type II fuzzy image segmentation, in *Fuzzy Sets and Their Extension: Representation, Aggregation and Models*, Studies in Fuzziness and Soft Computing, H. Bustince et al. (Eds.), Springer, Berlin, Heidelberg, 607–619, 2007.
24. Vlachos, I.K. and Sergiadis, G.D., Intuitionistic fuzzy information – Applications to pattern recognition, *Pattern Recognition Letters*, 28, 197–206, 2007.
25. Yasnoff, W.A., Mui, J.K., and Bacus, J.W., Error measures for scene segmentation, *Pattern Recognition*, 9, 217–231, 1977.

7

Clustering of Medical Images

7.1 Introduction

Segmentation is a key step towards image analysis in various image processing applications such as object recognition, pattern recognition and medical imaging. It can be defined as a grouping in a parameter space where points are associated with different sets of values of similar intensities in different images. So, grouping is the main step of image segmentation. This type of segmentation is called clustering which is very important in classifying different patterns/structures in an image. Clustering is a technique to separate unlabelled data into finite and discrete sets. It can be done using the fuzzy or non-fuzzy method. Traditional non-fuzzy clustering like k-means puts data into exactly one cluster. But for overlapped data sets where some data may be allocated to multiple clusters, k-means clustering may not analyse the data set clearly. To achieve better clustering, fuzzy c means clustering is used. The first fuzzy method to segment the regions of an image is the fuzzy c means (FCM) clustering, introduced by Bezdek et al. [2]. Clustering may be hard c means or FCM. Hard c means is a non-fuzzy method, which is also known as k means clustering. K means clustering partitions a collection of N vectors into k groups. It executes a sharp classification in which each object is assigned to a class or not. Also, there is very often no sharp boundary between clusters in many real-time images. This problem can be alleviated by associating a membership value in the interval [0, 1] to data in every cluster such that data that have a similarity with each cluster with membership values near 0 signify a small similarity between the sample and the cluster and data with membership values near 1 signify a high degree of similarity. Medical images contain a lot of uncertainties, and there are hardly sharp boundaries present and so fuzzy clustering may be very much beneficial. FCM partitions the data in such a way that a data point can belong to all groups in different membership grades where an element may have partial membership grades in several clusters – herein lies the distinction between the hard and fuzzy c partitions of data set X. It is an iterative algorithm where the aim is to find the cluster centres that minimize the dissimilarity function. This is an important feature in medical image diagnosis to increase the sensitivity.

7.2 Fuzzy c Means Clustering

The problem of fuzzy clustering may be viewed as an extension of crisp clustering. The algorithm requires a priori definition of a number of classes that will partition the image in different classes. It classifies the set of data points $X = (x_1, x_2, x_3, ..., x_n)$ into c homogeneous groups or clusters represented as fuzzy sets, $F = (F_1, F_2, F_3, ..., F_c)$.

Let $X = (x_1, x_2, ..., x_j, ..., x_n)$ be a set of sample, where data point x_k ($k = 1, 2, ..., n$) is required to be partitioned in $c(2 \leq c \leq n)$ clusters. Let us assume that u_{ik} is the membership grade of pattern x_k to the cluster i and $U = [u_{ik}]$ is a $c \times n$ membership matrix where u_{ik} is the membership grade of the ith object in the kth group:

$$U = \begin{array}{c} c_1 \\ c_2 \\ \vdots \\ c_n \end{array} \begin{bmatrix} u_{11} & u_{12} & \cdots & u_{1n} \\ u_{21} & & & \vdots \\ \vdots & & & \vdots \\ u_{c1} & \cdots & \cdots & u_{cn} \end{bmatrix}$$

The membership matrix implies that the nth data, x_n, belong to class c_1, $c_2, ..., c_c$, with membership functions $u_{1n}, u_{2n}, ..., u_{cn}$, respectively. The membership distribution has the following properties:

$$u_{ik} \in [0, 1], \quad \forall i, k\ i = 1, 2, 3, ..., n, k = \text{no. of classes}$$

$$0 < \sum_{i=1}^{n} u_{ik} < n \quad 1 \leq k \leq c$$

and

$$\sum_{k=1}^{c} u_{ik} = 1.0 \quad 1 \leq j \leq n$$

The objective is to obtain a c partition by minimizing the criterion function using the Lagrangian multiplier method:

$$J_m(U, V : X) = \sum_{i=1}^{n} \sum_{k=1}^{c} \mu_{ik}^m d^2(x_i, v_k) \tag{7.1}$$

The membership matrix U is randomly initialized as

$$u_{ik} = \frac{1}{\sum_{j=1}^{c} \left(\left(\|x_i - v_k\| \right) / \left(\|x_i - v_j\| \right) \right)^{2/(m-1)}}, \quad \forall k = 1, 2, \ldots, c, \quad i = 1, 2, \ldots, n \tag{7.2}$$

and the cluster centres are computed as

$$v_k = \frac{\sum_{i=1}^{n} u_{ik}^m x_i}{\sum_{i=1}^{n} u_{ik}^m} \tag{7.3}$$

where

u_{ik} is the membership of the data x_i to the kth fuzzy cluster with centroid v_k

m is user defined, and generally it is taken as 2

'd' is the Euclidean distance, or any distance measure is used to find the similarity between the cluster centre and the data points

This iteration will stop when $\max_{ik}\{|u_{ik}^{(t+1)} - u_{ik}^{(t)}|\} < \varepsilon$, where ε is a tolerance level which lies in between 0 and 1. t and $t + 1$ are the successive iterations.

An example of fuzzy clustering is shown in Figure 7.1.

Many studies are reported in the literature to improve the fuzzy clustering algorithm [10].

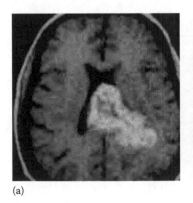 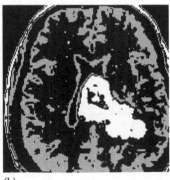

(a)   (b)

FIGURE 7.1
(a) Brain image and (b) fuzzy clustering into four regions.

7.3 Hierarchical Clustering

Hierarchical clustering is a method of cluster analysis that builds a hierarchy of clusters. There are two types of hierarchical clustering: agglomerative and divisive. The result of hierarchical clustering can be presented in a dendrogram as shown in Figure 7.2.

It shows how samples are grouped together. Agglomerative clustering merges clusters iteratively and follows a bottom-up approach. The procedure starts with each object as one cluster and forms a nested sequence by subsequently merging all the clusters. Grouping can be done on the basis of the nearest distance measure. It can be (1) a single linkage in which the minimum distance is computed between the clusters (one can set a threshold, and the clustering is stopped once the distance between the clusters is above the threshold); (2) a complete linkage where the maximum distance between the clusters is computed, that is, the farthest distance, and the algorithm stops when the maximum distance between the nearest clusters exceeds an arbitrary threshold and (3) an average that computes the average distance.

The main advantage of agglomerative clustering is that the cluster numbers are not fixed and no initialization and optimization criteria are required. Suppose $X = \{x_1, x_2, \ldots, x_n\}$ are samples for clustering. In level 1, samples are partitioned randomly into a finite set of p_0 partitions and all the samples are considered as single clusters. Then the least distance pair of clusters, say C_i and C_j, is computed as $d[C_i, C_j] = \min[d(x), d(y)]$, where the minimum is computed between all the pairs of clusters. The two clusters C_i and C_j are merged to a single cluster. In the next level, the distance matrix is updated by

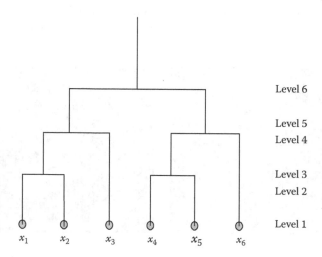

FIGURE 7.2
Dendrogram.

deleting the rows and column of the two clusters C_i and C_j that are merged, and a similar procedure is followed in the next level where the distance is computed between the old cluster, that is, the merged cluster (say C_m) and a new cluster (C_i, C_j). This labelling procedure continues till a single cluster is obtained. One may cut the dendrogram at any level to produce different clustering. In this method, no a priori information about the clusters is required. It is useful when there is a large amount of data.

There is also a *divisive* hierarchical clustering which does the reverse by starting with all objects in one cluster and subdividing them into smaller pieces. It follows a top-down approach, and the method starts with one cluster and the clusters are split recursively as one move downs the hierarchy. Divisive methods are not generally available and rarely have been applied. It starts with all the objects in one cluster. Clusters are subdivided into smaller clusters till each object forms a cluster on its own following the condition of termination criterion. A cluster splits according to the maximum Euclidian distance between the closest neighbouring objects in the cluster.

7.4 Kernel Clustering

To improve the FCM algorithm for accurate detection of boundaries, an alternative approach is used that transforms the input data to a high-dimensional feature space using a non-linear mapping function so that non-linearity in the input data can be treated linearly in the feature space according to Mercer's theorem. Direct computation in high-dimensional feature space consumes much time, and thus, Mercer kernels are used to make it practical. A popular transformation of data is the use of the kernel method. The inner product in the clustering algorithm is replaced by a kernel function. Clustering is performed after transforming the input data to a high-dimensional feature space.

Let us define a non-linear map $\phi : x \rightarrow \phi(x) \in F$, where $x \in X$, X denotes the data space and F is the transformed feature space. With the incorporation of kernel, the objective function in the feature space is rewritten as

$$J_m(U,V) = \sum_{i=1}^{n} \sum_{k=1}^{c} u_{ik}^m \left\| \phi(x_i) - \phi(v_k) \right\|^2 \tag{7.4}$$

where

$$\left\| \phi(x_i) - \phi(v_k) \right\|^2 = \left\langle \phi(x_i), \phi(x_i) \right\rangle + \left\langle \phi(v_k), \phi(v_k) \right\rangle - 2 \left\langle \phi(x_i), \phi(v_k) \right\rangle \tag{7.5}$$

Expressing $K(x_i, v_k) = \left\langle \phi(x_i), \phi(v_k) \right\rangle$

The new kernel function distance function is written as

$$\left\|\varphi(x_i) - \varphi(v_k)\right\|^2 = K(x_i, x_i) + K(v_k, v_k) - 2K(x_i, v_k) \tag{7.6}$$

where
$i = 1, 2, \ldots, n$
$k = 1, 2, \ldots, c$

The kernel function may be a radial basis function (RBF), hypertangent, Gaussian or polynomial kernel.

$$\text{Hypertangent kernel} \quad H(x_i, v_k) = 1 - \tanh\left(\frac{-\left\|x_i - v_k\right\|^2}{\sigma^2}\right)$$

Using this function, the distance function becomes

$$\left\|\varphi(x_i) - \varphi(v_k)\right\|^2 = H(x_i, x_i) + H(v_k, v_k) - 2H(x_i, v_k) \quad \text{as } H(x_i, x_i) = 1, H(v_k, v_k) = 1$$

so,

$$\left\|\varphi(x_i) - \varphi(v_k)\right\|^2 = 2(1 - H(x_i, v_k)) \tag{7.7}$$

$$\text{Gaussian kernel} \quad G(x_i, v_k) = \exp\left(\frac{-\left\|x_i - v_k\right\|}{\sigma^2}\right) \tag{7.8}$$

In this case, $G(x_i, x_i) = 1$, $G(v_k, v_k) = 1$, so $||\varphi(x_i) - \varphi(v_k)||^2 = 2(1 - G(x_i, v_k))$:

$$\text{Radial basis kernel} \quad R(x_i, v_k) = \exp\left(-\frac{\left|x_i^a - v_k^a\right|^b}{\sigma^2}\right)$$

where σ, a and b are the adjustable parameters.
   In this case also,

$$\left\|\varphi(x_i) - \varphi(v_k)\right\|^2 = 2(1 - R(x_i, v_k)) \tag{7.9}$$

To show the application of kernel clustering on images, a few examples are given.

7.5 Kernel Clustering Methods

1. Kannan et al. [11] suggested a robust FCM algorithm to segment medical images. They modified the objective function where they replaced the Euclidean distance by a hypertangent kernel function. The objective function as in Equation 7.4 is

$$J_m(U,V) = \sum_{i=1}^{n}\sum_{k=1}^{c} u_{ik}^m \left\| \varphi(x_i) - \varphi(v_k) \right\|^2$$

For hypertangent kernel

$$K(x_i, v_k) = 1 - \tanh\left(\frac{-\left\| x_i - v_k \right\|^2}{\sigma^2} \right)$$

where σ is a parameter adjusted by the users.
The objective function using Equation 7.7 reduces to

$$J_m(U,V) = 2\sum_{i=1}^{n}\sum_{k=1}^{c} u_{ik}^m (1 - K(x_i, v_k))$$

Using Lagrangian multiplier, the objective function can be written as

$$J_m(U,V) = 2\sum_{i=1}^{n}\sum_{k=1}^{c} u_{ik}^m (1 - K(x_i, v_k)) - \sum_{i=1}^{n}\lambda_i\left(\sum_{k=1}^{c} u_{ik} - 1 \right) \tag{7.10}$$

Expanding this equation,

$$J_m(U,V) = 2\left[\sum_{i=1}^{n} u_{i1}^m (1 - K(x_i, v_1)) + \cdots + u_{ic}^m (1 - K(x_i, v_c)) \right]$$

$$- \sum_{i=1}^{n}\lambda_i ((u_{i1} + u_{i2} + \cdots + u_{ic}) - 1)$$

Taking the derivative of J with respect to u and setting the first derivative to zero, we get

$$\frac{\delta J}{\delta u_{11}} = 2m u_{11}^{m-1}(1 - K(x_1, v_1)) - \lambda_1 = 0$$

$$u_{11} = \left(\frac{\lambda_1}{2m(1 - K(x_1, v_1))}\right)^{1/(m-1)} \tag{7.11}$$

Likewise, for

$$u_{12} = \left(\frac{\lambda_1}{2m(1 - K(x_1, v_2))}\right)^{1/(m-1)}$$

$$\vdots$$

$$u_{1c} = \left(\frac{\lambda_1}{2m(1 - K(x_1, v_c))}\right)^{1/(m-1)}$$

Summing up all u's we get

$$u_{11} + u_{12} + \cdots + u_{1c} = \left(\frac{\lambda_1}{2m(1 - K(x_1, v_1))}\right)^{1/(m-1)} + \left(\frac{\lambda_1}{2m(1 - K(x_1, v_2))}\right)^{1/(m-1)}$$

$$+ \cdots + \left(\frac{\lambda_1}{2m(1 - K(x_1, v_c))}\right)^{1/(m-1)}$$

$$= \left(\frac{\lambda_1}{2m}\right)^{1/(m-1)} \left[\sum_{j=1}^{c} \frac{1}{(1 - K(x_1 - v_j))}\right]^{1/(m-1)} \tag{7.12}$$

As per the objective criterion, $\sum_{k=1}^{c} \mu_{ik} = 1$.
   So,

$$\left(\frac{\lambda_1}{2m}\right)^{1/(m-1)} \left[\sum_{j=1}^{c} \frac{1}{(1 - K(x_1 - v_j))}\right]^{1/(m-1)} = 1$$

$$\left(\frac{\lambda_1}{2m}\right)^{1/(m-1)} = \frac{1}{\sum_{j=1}^{c} (1/(1 - K(x_1 - v_j)))^{1/(m-1)}} \tag{7.13}$$

Substituting Equation 7.13 in u_{11} in Equation 7.11, we have

$$u_{11} = \frac{\left(1 - K(x_1, v_1)\right)^{-1/(m-1)}}{\sum_{j=1}^{c} \left(1 - K(x_1, v_j)\right)^{-1/(m-1)}}$$

In general,

$$u_{ik} = \frac{\left(1 - K(x_i, v_k)\right)^{-1/(m-1)}}{\sum_{j=1}^{c} \left(1 - K(x_i, v_j)\right)^{-1/(m-1)}} \qquad (7.14)$$

To compute the cluster centre, the equation for the objective function is

$$J_m(U,V) = 2 \sum_{i=1}^{n} \sum_{k=1}^{c} u_{ik}^m \left(\tanh\left(\frac{-\|x_i - v_k\|^2}{\sigma^2} \right) \right) - \sum_{i=1}^{n} \lambda_i \left(\sum_{k=1}^{c} u_{ik} - 1 \right) \qquad (7.15)$$

Expanding this equation and taking the first derivative with respect to v and equating to zero, we obtain

$$\frac{\delta J}{\delta v_1} = 2 \sum_{i=1}^{n} u_{i1}^m \left(1 - \tanh^2 \frac{-\|x_i - v_1\|^2}{\sigma^2} \right) \frac{(x_i - v_1)}{\sigma^2} = 0 \quad \text{as } d/dx \,(\tanh x)$$

$$= 1 - \tanh^2 x \quad \text{(or)}$$

$$\sum_{i=1}^{n} u_{i1}^m \left(1 + \tanh \frac{-\|x_i - v_1\|^2}{\sigma^2} \right) \left(1 - \tanh \frac{-\|x_i - v_1\|^2}{\sigma^2} \right) (x_i - v_1) = 0 \quad \text{(or)}$$

$$\sum_{i=1}^{n} u_{i1}^m \left(1 + \tanh \frac{-\|x_i - v_1\|^2}{\sigma^2} \right) H(x_i, v_1)(x_i - v_1) = 0 \quad \text{(or)}$$

$$\sum_{i=1}^{n} u_{i1}^m H(x_i, v_1) \left(1 + \tanh \frac{-\|x_i - v_1\|^2}{\sigma^2} \right) x_i$$

$$- \sum_{i=1}^{n} u_{i1}^m H(x_i, v_1) \left(1 + \tanh \frac{-\|x_i - v_1\|^2}{\sigma^2} \right) v_1 = 0$$

$$v_1 = \frac{\sum_{i=1}^{n} u_{i1}^m H(x_i, v_1)\left(1 + \tanh\frac{-\|x_i - v_1\|^2}{\sigma^2}\right)x_i}{\sum_{i=1}^{n} u_{i1}^m H(x_i, v_1)\left(1 + \tanh\frac{-\|x_i - v_1\|^2}{\sigma^2}\right)} \tag{7.16}$$

Likewise, for v_k,

$$v_k = \frac{\sum_{i=1}^{n} u_{ik}^m H(x_i, v_k)\left(1 + \tanh\frac{-\|x_i - v_k\|^2}{\sigma^2}\right)x_i}{\sum_{i=1}^{n} u_{ik}^m H(x_i, v_k)\left(1 + \tanh\frac{-\|x_i - v_k\|^2}{\sigma^2}\right)}$$

Following a similar procedure as FCM, iteration will stop when $\{|u_{ik}^{(t+1)} - u_{ik}^{(t)}|\} < \varepsilon$, where ε is a tolerance level which lies in between 0 and 1. t and $t + 1$ are successive iterations.

Example 7.1

Two examples on computed tomography (CT) scan images of the brain are shown to show the effectiveness of the kernel function in the original FCM algorithm (Figure 7.3).

2. To increase the robustness of FCM, Ahmed et al. [1] considered spatial neighbourhood information and the objective function is modified as

$$J_m(U,V) = \sum_{i=1}^{n}\sum_{k=1}^{c} u_{iik}^m \|x_i - v_k\|^2 + \frac{\alpha}{N_R}\sum_{i=1}^{n}\sum_{k=1}^{c} u_{ik}^m \sum_{x_r \in N_i} \|x_r - v_k\|^2$$

where
N_i is the set of pixel neighbours in a window around x_i
N_R is the cardinality of N_i
α is the controlling parameter of the neighbouring terms
$\bar{x}_i$ is the sample mean within the window in the neighbourhood of x_i

The membership and cluster centres are updated as

$$u_{ik} = \frac{\left(\|x_i - v_k\|^2 + (\alpha/N_R)\sum_{x_r \in N_i}\|x_r - v_k\|^2\right)^{-1/(m-1)}}{\sum_{j=1}^{c}\left(\|x_i - v_j\|^2 + (\alpha/N_R)\sum_{x_r \in N_i}\|x_r - v_k\|^2\right)^{-1/(m-1)}}$$

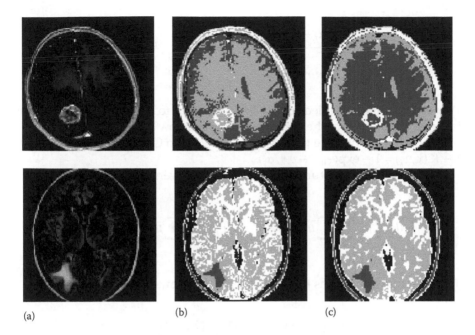

(a) (b) (c)

FIGURE 7.3
(a) CT scan 'clot 1' image, (b) FCM cluster and (c) FCM modified with kernel.

$$v_k = \frac{\sum_{i=1}^{n} u_{ik}^m \left(x_i + (\alpha/N_R) \sum_{x \in N_i} x_r \right)}{(1+\alpha) \sum_{i=1}^{n} u_{ik}^m}, \quad i = 1, 2, 3, \ldots, n, \quad k = 1, 2, 3, \ldots, c$$

3. Chen and Zhang [8,15] pointed that in Ahmed's method, the neighbourhood term N_i takes a lot of time than FCM. So, they modified the objective function by removing the term N_i as

$$J_m(U,V) = \sum_{i=1}^{n} \sum_{k=1}^{c} u_{ik}^m \|x_i - v_k\|^2 + \alpha \sum_{i=1}^{n} \sum_{k=1}^{c} u_{ik}^m \|\bar{x}_i - v_k\|^2 \qquad (7.17)$$

The effect of neighbouring terms is controlled by α, and it lies $1 \leq \alpha \leq 0$. The necessary condition on u_{ik} to be at local minimum is

$$u_{ik} = \frac{\left(\|x_i - v_k\|^2 \right) + \alpha \|\bar{x}_i - v_k\|^2 \right)^{-1/m-1}}{\sum_{j=1}^{c} \left(\|x_i - v_j\|^2 + \alpha \|\bar{x}_i - v_k\|^2 \right)^{-1/m-1}}$$

$$v_k = \frac{\sum_{i=1}^{n} u_{ik}^m \left(x_i + \alpha \overline{x}_i\right)}{(1+\alpha)\sum_{i=1}^{n} u_{ik}^m}, \quad i=1,2,3,\ldots,n, \quad k=1,2,3,\ldots,c$$

When $\alpha = 0$, the modified algorithm converges to the original FCM. $\overline{x}_i$ is taken as the median of the neighbours in a window around x_i.

The Euclidean distance $\|x_i - v_k\|$ is replaced with Gaussian kernel distance $1 - K(x_i, v_k) = 1 - \exp(-\|x_i - v_k\|/\sigma^2)$.

So, the new objective function with the kernel version is written as

$$J_m(U,V) = \sum_{k=1}^{c}\sum_{i=1}^{n} u_{ik}^m \left(1 - K(x_i, v_k)\right) + \alpha \sum_{k=1}^{c}\sum_{i=1}^{n} u_{ik}^m \left(1 - K(\overline{x}_i, v_k)\right) \quad (7.18)$$

The necessary conditions to minimize are as follows:

$$u_{ik} = \frac{\left(1 - K(x_i, v_k) + \alpha(1 - K(\overline{x}_i, v_k))\right)^{-1/(m-1)}}{\sum_{j=1}^{c}\left((1 - K(x_i, v_j)) + \alpha(1 - K(\overline{x}_i, v_j))\right)^{-1/(m-1)}}$$

$$v_k = \frac{\sum_{i=1}^{n} \mu_{ik}^m K(x_i, v_k) x_i + \alpha K(\overline{x}_i, v_k)\overline{x}_i}{\sum_{i=1}^{n} \mu_{ik}^m (K(x_i, v_k) x_i + \alpha K(\overline{x}_i, v_k))}$$

The usual clustering procedure then follows.

Example 7.2

An example is shown on CT scan images of the brain about the efficacy of the kernel function in the original FCM algorithm along with spatial information (Figure 7.4).

It is observed that fuzzy clustering with kernel can detect the blood clot clearly as the kernel function transforms the input image to a high-dimensional feature space. Now, the effect of intuitionistic fuzzy set on medical image clustering is shown where more uncertainties are considered. It may be useful in those images where the fuzzy method may not give better results.

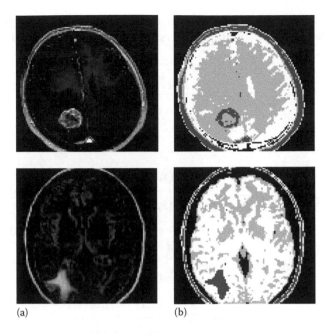

(a)   (b)

FIGURE 7.4
(a) CT scan brain image and (b) kernel clustered image.

7.6 Intuitionistic Fuzzy *c* Means Clustering

Uncertainty is largely present in medical images because of noise in image acquisition and low illumination. This is the reason why the regions or boundaries are not clear. Fuzzy set theory works well on these images as it considers imprecise information. But the intuitionistic fuzzy set theory considers more information compared to fuzzy set. It considers both the membership and non-membership degrees. As it considers more uncertainties, intuitionistic fuzzy set is assumed to work better on medical images. Medical images are very difficult to analyse, and there is very little work on clustering using intuitionistic fuzzy set on medical images. It follows a similar type of algorithm as FCM but in an intuitionistic way. To see how the intuitionistic property is incorporated into clustering, some works suggested by different authors are described in this section:

1. Iakovidis et al. [9] suggested a new similarity metric in conventional clustering algorithm where both membership and non-membership degrees are included. The similarity metric is based on the histogram intersection technique. Let $A(\mu_A, v_A)$ and $B(\mu_B, v_B)$ be two intuitionistic fuzzy sets where

μ_A, μ_B and v_A, v_B are the membership and non-membership degrees, respectively. The intuitionistic fuzzy intersection similarity between two sets A and B is written as

$$S_{IFS}(A,B) = \frac{s_1(\mu_A,\mu_B) + s_1(v_A,v_B)}{2} \tag{7.19}$$

where

$$s_1(\mu_A,\mu_B) = \begin{cases} \dfrac{\sum_{i=1}^n \min(\mu_A(x_i),\mu_B(x_i))}{\sum_{i=1}^n \max(\mu_A(x_i),\mu_B(x_i))} & \text{if } \mu_A \cup \mu_B \neq 0 \\ 1 & \text{if } \mu_A \cup \mu_B = 0 \end{cases}$$

and for the non-membership degree,

$$s_1(v_A,v_B) = \frac{\sum_{i=1}^n \min(v_A(x_i),v_B(x_i))}{\sum_{i=1}^n \max(v_A(x_i),v_B(x_i))}$$

where $\mu_A = \{\mu_A(x)\}$ and $v_A = \{v_A(x)\}$.

The new objective criterion is reformulated as

$$J_m^{IFS}(U,V:X) = \sum_{i=1}^n \sum_{k=1}^c \mu_{ik}^m |x_i - v_k|_{IFS}, \quad \text{with } 0 < \sum_{i=1}^n \mu_{ik} < n, \; \sum_{k=1}^c \mu_{ik} = 1$$

$|x_i - v_k|_{IFS} = 1 - S_{IFS}$ is the distance between the data vector x_i and cluster centre v_k. On minimizing $J_m^{IFS}(U,V:X)$, we get as in FCM

$$u_{ik} = \frac{\left(|x_i - v_k|_{IFS}\right)^{-1/(m-1)}}{\sum_{j=1}^c \left(|x_i - v_j|_{IFS}\right)^{-1/(m-1)}}, \quad \forall u_{ik}, 1 < k < c, \; 1 < i < N$$

And the centroid is computed as

$$v_k = \frac{\sum_{i=1}^n u_{ik}^m x_i}{\sum_{i=1}^n u_{ik}^m}$$

The distance function solely depends on the membership and non-membership values after the computation of centroids and before the next iterations.

Example 7.3

An example of a brain clot/haemorrhage image is shown. It is observed that the clustered image without kernel using intuitionistic fuzzy set does not cluster image clearly as it contains noise, but it is better than fuzzy clustering (Figure 7.5).

2. Chaira [3,4] suggested another intuitionistic clustering algorithm. In the construction of the algorithm, the following steps are required: (a) creating an intuitionistic fuzzy set using Yager-type intuitionistic fuzzy generator and (b) modifying the objective criterion function, updating the cluster centre using intuitionistic fuzzy set and incorporating an objective function, that is, the intuitionistic fuzzy entropy (*IFE*), in the criterion function.

An objective function, that is, *IFE*, is introduced in the conventional clustering algorithm. The *IFE* by Chaira [4] is given as (described in Chapter 4)

$$IFE(A) = \sum_{i=1}^{n} \pi_A(x_i) \cdot e^{[1-\pi_A(x_i)]} \qquad (7.20)$$

where $\pi_A(x_i) = 1 - (\mu_A(x_i) + \nu_A(x_i))$.

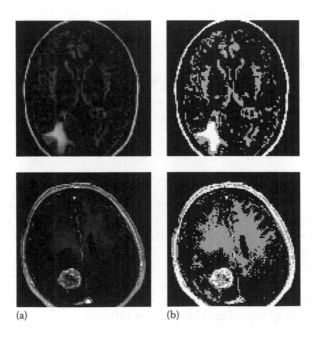

(a)   (b)

FIGURE 7.5
(a) CT scan brain image and (b) intuitionistic fuzzy method by Dimitris.

Intuitionistic fuzzy set is constructed using Yager's intuitionistic fuzzy complement and is written as

$$A_\lambda^{IFS} = \left\{ x, \mu_A(x), (1-\mu_A(x)^\alpha)^{1/\alpha} \mid x \in X \right\}$$

and the hesitation degree is

$$\pi_A(x) = 1 - \mu_A(x) - (1 - \mu_A(x)^\alpha)^{1/\alpha} \tag{7.21}$$

In this intuitionistic fuzzy clustering algorithm, the criterion function contains two terms: (a) the intuitionistic-type objective function as in conventional FCM and (b) the *IFE*.

A second objective function is introduced which is the *IFE* that aims in maximizing the good points in the class. The goal is to minimize the entropy of the histogram of an image. The hesitation values of all the elements in each cluster are added, and then the entropy of each class is calculated. It tells the amount of fuzziness or uncertainty present in the cluster. The second function is

$$J_2 = \sum_{k=1}^{c} \pi_k^* e^{1-\pi_k^*}$$

where $\pi_k^* = (1/N) \sum_{i=1}^{n} \pi_{ik}$, $k \in [1, N]$. π_{ik} is the hesitation degree of the *i*th element in cluster *k*.

So, the final criterion function that contains two terms is minimized and is as follows:

$$J = \sum_{i=1}^{n} \sum_{k=1}^{c} u_{ik}^{*m} d(x_i, v_k)^2 + \sum_{k=1}^{c} \pi_k^* e^{1-\pi_k^*}, \quad \text{with } m = 2$$

where

$d(x_i, v_k)$ is the Euclidean distance measure (or any distance measure) between v_k (cluster centre) of each region and x_i (points in the pattern)
u_{ik} is the membership value of the *i*th data (x_i) in the *k*th cluster
c is the number of clusters
n is the number of data points

The hesitation degree is calculated using Equation 7.21, and the intuitionistic fuzzy membership values are obtained as follows:

$$u_{ik}^* = u_{ik} + \pi_{ik} \tag{7.22}$$

where $u_{ik}^*(u_{ik})$ denotes the intuitionistic (conventional) fuzzy membership matrix of the ith data in kth class.

Substituting Equation 7.22 in the conventional FCM method, the modified cluster centre is written as

$$v_k^* = \frac{\sum_{i=1}^{n} u_{ik}^* x_i}{\sum_{i=1}^{n} u_{ik}^*} \tag{7.23}$$

Using this equation, the cluster centre is updated and simultaneously the membership matrix is updated.

At each iteration, the cluster centre and the membership matrix are updated and the algorithm stops when the difference of the updated membership matrix and the previous matrix is less than ε, that is,

$$\max_{i,k} \left| U_{ik}^{*\text{new}} - U_{ik}^{*\text{prev}} \right| < \varepsilon$$

ε is a user-defined value and is selected as $\varepsilon = 0.03$.

Thus, the criterion function in conventional FCM is modified using intuitionistic fuzzy set.

In the clustering algorithm, three features are considered, namely, pixel grey value, pixel mean and standard deviation. A small square window of size 3×3 is used throughout the image to calculate the mean and the standard deviation. Regarding the selection of α in Equation 7.21, with $\alpha \leq 0.5$, images are binary thresholded, and with $\alpha > 0.5$, clustered images are obtained. But better results are obtained with $\alpha = 0.85$.

Example 7.4

An example is shown on CT scan images of the brain to show the efficacy of the algorithm. Figure 7.6 shows the tumour/clot in the brain. Along with the intuitionistic fuzzy method, conventional FCM algorithm is also shown.

The intuitionistic fuzzy clustering method clusters the tumour from the image clearly.

7.7 Kernel-Based Intuitionistic Fuzzy Clustering

In intuitionistic fuzzy c means (IFCM) clustering, all the regions are not clustered clearly and it is not robust to noise. To overcome the drawback of intuitionistic fuzzy clustering and make it more robust, the objective function

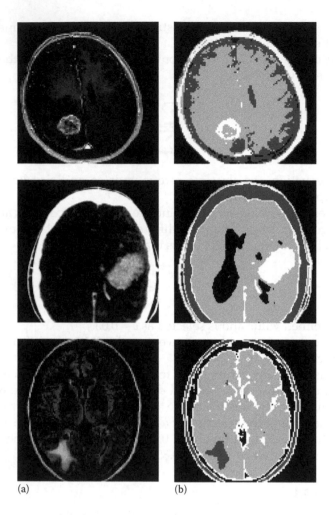

(a)   (b)

FIGURE 7.6
(a) Brain image 'tumour' and (b) intuitionistic fuzzy cluster. (Modified from Chaira, T. and Panwar, A., *J. Comput. Intell. Syst.*, 7(2), 1–11, 2013.)

in the IFCM algorithm is modified by Chaira and Panwar [7] using a kernel distance function in the place of Euclidean distance. The objective function as in Equation 7.4 is

$$J_m(U,V) = \sum_{i=1}^{n}\sum_{k=1}^{c} u_{ik}^m \left\| \varphi(x_i) - \varphi(v_k) \right\|^2$$

where from Equation 7.6

$$\|\varphi(x_i)-\varphi(v_k)\|^2 = K(x_i,x_i)+K(v_k,v_k)-2K(x_i,v_k) \quad i=1,2,\ldots,n \text{ and } k=1,2,\ldots,c$$

where
  i is the data point
  k is the cluster centre

Three kernels, namely, the hypertangent kernel, Gaussian kernel and radial basis kernel are used.
  For all these kernels, $K(x_i, x_i) = 1$ and $K(v_k, v_k) = 1$.
  The objective function in FCM reduces to

$$J_m(U,V)=2\sum_{i=1}^{n}\sum_{k=1}^{c} u_{ik}^m (1-K(x_i,v_k))$$

The membership matrix is

$$u_{ik}=\frac{\left(1-K(x_i,v_k)\right)^{-1/(m-1)}}{\sum_{j=1}^{c}\left(1-K(x_i,v_j)\right)^{-1/(m-1)}}$$

The modified intuitionistic fuzzy membership degree is $u_{ik}^* = u_{ik} + \pi_{ik}$; u_{ik} is the original membership matrix in FCM where $\pi_{ik} = 1 - u_{ik} - (1 - u_{ik})/(1 + \lambda \cdot u_{ik})$, and λ is taken as 1. Sugeno's fuzzy complement is used to compute the non-membership degree.
  The cluster centre after incorporating the intuitionistic property using hypertangent kernel is written as

$$v_i=\frac{\sum_{k=1}^{n}\mu_{ik}^{*m}K(x_i,v_k)\left(1+\tanh\left(\frac{-\|x_k-v_i\|^2}{\sigma^2}\right)\right)x_i}{\sum_{k=1}^{n}\mu_{ik}^{*m}K(x_i,v_k)\left(1+\tanh\left(\frac{-\|x_k-v_i\|^2}{\sigma^2}\right)\right)}$$

The cluster centre using Gaussian and radial basis kernel is written as

$$v_i=\frac{\sum_{k=1}^{n}\mu_{ik}^{*m}K(x_i,v_k)x_i}{\sum_{k=1}^{n}\mu_{ik}^{*m}K(x_i,v_k)} \tag{7.24}$$

The membership matrix will follow Equation 7.14 using hypertangent kernel, but hesitation degree is incorporated into each membership degree.

Kernel-based IFCM clustering is robust on noise. It works well in a noisy environment. Examples are shown on tumour images using all the three kernels.

Example 7.5

Two examples are shown on CT scan images of the brain to show the efficacy of the algorithm. Figures 7.7 and 7.8 show the tumour/clot in the brain.

It is observed that simple intuitionistic fuzzy clustering does not perform better in a noisy environment but works well in a noisy environment. Radial basis kernel does not work well. Hypertangent kernel works a little bit better than Gaussian kernel where the tumour/clot is clearly detected.

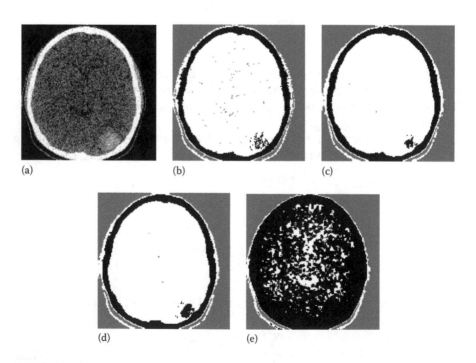

FIGURE 7.7
(a) Brain clot image, (b) IFCM, (c) IFCM with Gaussian kernel, (d) IFCM with hypertangent kernel and (e) IFCM with radial basis kernel. (Modified from Chaira, T. and Panwar, A., *J. Comput. Intell. Syst.*, 7(2), 1–11, 2013.)

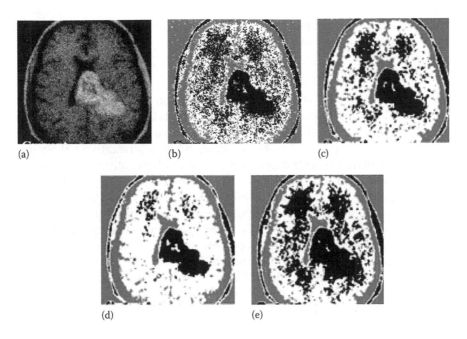

(a)   (b)   (c)

(d)   (e)

FIGURE 7.8
(a) Original image, (b) IFCM, (c) IFCM with Gaussian kernel, (d) IFCM with hypertangent kernel and (e) IFCM with radial basis kernel. (Modified from Chaira, T. and Panwar, A., *J. Comput. Intell. Syst.*, 7(2), 1–11, 2013.)

7.8 Colour Clustering

The objective of colour clustering is to divide or cluster the image into several homogeneous regions using colour as a feature. Colour approach is an important issue especially when we are dealing with medical images, for example, cells or tissues or skin-related problems, where the effect of colour clustering is analysed to diagnose different types of diseases. Recently, cell classification has a widespread interest especially in pathological laboratories where blood cell counting of red blood cell (RBC) and white blood cell (WBC) is carried out for disease detection. Blood cell can be RBC, WBC and blood platelets, and separating WBC from the blood cell is a very critical task. The difference in these types lies in the texture, colour of the cytoplasm and nucleus. In blood smear images, the RBCs are more than the WBCs. Accurate cell image segmentation for determining valuable quantitative diagnostic information for pathologists is necessary. Now, with the automated techniques such as medical imaging, analysing becomes faster and gives more accurate results. But the requirements for clustering pathological images are different from the general clustering of images as the colours (stained cell images) are vaguely distributed and so clustering of cells becomes very difficult. Generally, unsupervised method is used to cluster the images that automatically find the structures in the data set.

7.8.1 Colour Model

Colour models are the mathematical representation of a set of colours. There are different types of colour models such as RGB (red, green, blue), HSV (hue, saturation, value) and CIELab. In CIELab, the three coordinates are L^*, a^*, b^*: $L^* = 0$ indicates black and $L^* = 100$ indicates white, a^*- negative values indicate green, positive values indicate magenta, and b^*- negative values indicate blue, positive values indicate yellow. RGB is the common representation of colour image due to its simple representation. It is device dependent and is not a perceptual model. CIELab colour space is a perceptually human model where the Euclidean distance between two colour points is equivalent to the difference between two colours as perceived by humans. It is denoted by three Cartesian coordinates: L^*, a^* and b^*. To obtain a CIELab colour model from RGB, two steps are required: First, it transforms the RGB to CIE XYZ colour model and then CIE XYZ to CIELab using the following formulas [12].

Transformation of RGB to CIE XYZ

$$X = 0.49R + 0.310G + 0.200B$$

$$Y = 0.177R + 0.813G + 0.010B$$

$$Z = 0.000R + 0.010G + 0.990B$$

Transformation of CIE XYZ to CIELab

$$L^* = \begin{cases} 116\left(\dfrac{Y}{Y_N}\right)^{(1/3)} - 16 & \text{for } \left(\dfrac{Y}{Y_N}\right) > 0.008856 \\[2ex] 903.3\left(\dfrac{Y}{Y_N}\right) & \text{for } \left(\dfrac{Y}{Y_N}\right) \leq 0.008856 \end{cases}$$

$$a^* = 500\left[f\left(\frac{X}{X_N}\right) - f\left(\frac{Y}{Y_N}\right) \right]$$

$$b^* = 200\left[f\left(\frac{Y}{Y_N}\right) - f\left(\frac{Z}{Z_N}\right) \right]$$

where

$$f(t) = \begin{cases} t^{(1/3)}, & t > 0.008856 \\[2ex] 7.787t + \dfrac{16}{116}, & \text{otherwise} \end{cases}$$

where X_N, Y_N and Z_N are the CIE tristimulus values with reference to the white point (D_{65}).

The colour clustering algorithm, suggested by Chaira [5,6] using intuition-istic fuzzy set theory, is used to segment human cell images that use CIELab colour space. This method clusters the nucleus and cytoplasm of pathologi-cal RBC or WBC in the blood cell image. Other colour spaces such as RGB and HSV are also used. But CIELab colour space gives better results as it is a perceptually dependent colour model.

Initially, Sugeno-type intuitionistic fuzzy complement is used to construct an intuitionistic fuzzy set. Thus, with the help of Sugeno fuzzy generator, *IFS* becomes

$$A_\lambda^{IFS} = \left\{ x, \mu_A(x), \left. \frac{(1-\mu_A(x))}{(1+\lambda \cdot \mu_A(x))} \right| x \in X \right\}$$

with hesitation degree

$$\pi_A(x) = 1 - \mu_A(x) - \frac{(1-\mu_A(x))}{(1+\lambda \cdot \mu_A(x))} \qquad (7.25)$$

Six features for each data point are taken – 3 for L, a and b values of each colour pixel and 3 for the mean of L, a and b values of the neighbourhood of the pixel. For each pixel, the mean or the average values of L, a and b are cal-culated in the 3×3 window neighbourhood. This is moved throughout the image to find the average for each pixel neighbourhood, which is similar to the conventional FCM clustering. The value 'λ' in Equation 7.25 plays a signif-icant role in intuitionistic fuzzy clustering. When the value of λ is increased gradually from 2, the clustered image is degraded, that is, the regions are not properly clustered. With the value of $\lambda = 1$, the cluster seems to be better.

Performance evaluation: In order to verify the performance of segmentation methods, ground truth or manually segmented images are constructed. It is similar to that of monochrome image performance computation. The mis-classification error of the clustered images is calculated as [14]

$$\text{Error} = 1 - \frac{\left| R_{ET}^1 \cap R_{GT}^1 \right| + \left| R_{ET}^2 \cap R_{GT}^2 \right| + \left| R_{ET}^3 \cap R_{GT}^3 \right| + \left| R_{ET}^4 \cap R_{GT}^4 \right|}{R_{GT}^1 + R_{GT}^2 + R_{GT}^3 + R_{GT}^4}$$

where

$R_{ET}^1, R_{ET}^2, R_{ET}^3$ and R_{ET}^4 are the regions in the experimental segmented/clustered image

$R_{GT}^1, R_{GT}^2, R_{GT}^3$ and R_{GT}^4 are the different regions in the ground truth segmented/clustered image

This error reflects the percentage of wrongly assigned pixels that ranges from zero or no error, that is, when the image is exactly clustered, to 1, that is, when the image is wrongly clustered.

Example 7.6

Two examples are shown that are related to intuitionistic fuzzy colour clustering. Images are downloaded from the image gallery http://www.polconsultant.com/conteduc/hematology/micro/. To deal with cell images, staining is done to enhance the contrast and highlight the structures of cells or tissues for viewing with the help of a microscope. But in the figures, the images are converted to greyscale images using the code 'colormap (grey)'.

Figure 7.9 is an abnormal RBC image containing tooth-like cells called poikilocyte or burr cells. Figure 7.10 is a stained cell image showing the monocyte in purple colour. The results are displayed in black and white.

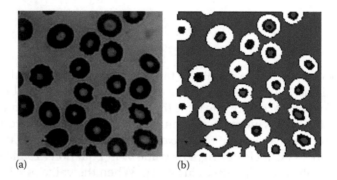

(a)   (b)

FIGURE 7.9
(a) Abnormal RBC image and (b) intuitionistic fuzzy cluster.

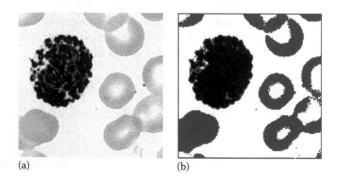

(a)   (b)

FIGURE 7.10
(a) Basophil image and (b) intuitionistic fuzzy cluster.

7.9 Type II Fuzzy Clustering

Clustering using Type II fuzzy on medical images is discussed in this section. There is no work based on medical image clustering using Type II fuzzy set. Rhee and Hwang [13] suggested a clustering algorithm, but the algorithm was tried on patterns.

Rhee and Hwang proposed type II fuzzy clustering of pattern set. Type II fuzzy set is the fuzziness in a fuzzy set. In this algorithm, the membership value of each pattern in the image is extended to type II fuzzy membership by assigning membership grades to Type I fuzzy membership. Any type of membership function may be used. If the membership value of the pattern is high, it is considered to have less uncertainty and vice versa. So, the higher the membership value of the pattern, the more is the contribution of the pattern to the cluster. On using type 2 membership, the contribution of the pattern that has a low membership in Type I will have a relatively lower membership which helps in representing the membership in a better way. In their work, they used a triangular membership function to obtain Type I membership. In doing so, cluster centres converge to a desirable location than cluster centres obtained in Type I membership. The membership values for type II membership are obtained as

$$a_{ik} = u_{ik} - \frac{1 - u_{ik}}{2}$$

where a_{ik} and u_{ik} are the type II and Type I fuzzy membership, respectively. The cluster centres are updated accordingly using conventional FCM taking into account the new type II fuzzy membership as

$$v_{ik}^{type2} = \frac{\sum_{k=1}^{n} a_{ik}^m x_{ik}}{\sum_{k=1}^{n} a_{ik}^m}$$

Then the usual procedure as in FCM follows. When performing clustering in a fuzzy set, the membership assigns the availability of the pattern in the clusters, but when applying the fuzzy membership to the pattern set, imperfect information lies in various parameters in the fuzzy membership assignment.

Supervised methods have been largely employed in medical image segmentation, but they require conditions, which are difficult to satisfy in clinical fields: (1) they require labelling a set of prototypical samples in order to apply the process of generalization; (2) if the number of clusters is defined, labelling of voxels in a training set belonging with certainty in different clusters is not trivial, especially when it contains multimodal data; (3) users have to introduce bias in consequence of the large inter-user variability, which is generally observed when manual labelling is performed and (4) it is time consuming especially for large volumes.

In unsupervised clustering, no user's definition of training samples is required and multidimensional data are clearly exploited.

A MATLAB® program for fuzzy clustering is given.

An image is clustered into four groups, and pixel is used as feature. An FCM clustering algorithm is used.

```
clear all;
image2=imread('tumor_2.jpg');imagergb=rgb2gray(image2);
dim=75;
image=imcrop(imagergb,[1 1 dim-1 dim-1]);
im1=double(image);
imf = im1(:);
[centerL,U,obj_fcn]=fcm(imf,4); % U = membership matrix,
center = cluster center, obj_fcn is the objective function,
4 is no. of clusters
maxU = max(U);
index1 = find(U(1,:)==maxU);
index2 = find(U(2,:)==maxU);
index3 = find(U(3,:)==maxU);
index4 = find(U(4,:)==maxU);
imf(index1)=0.5*ones;
imf(index2)=zeros;
imf(index3)=ones;
imf(index4)=0.7*ones;
imff=reshape(imf,[dim dim]);
figure,imshow (imff,[]);
```

7.10 Summary

This chapter presents in detail FCM clustering and hierarchical clustering. Kernel clustering is also discussed as it is robust to noise. The kernel function uses Gaussian or hypertangent or the radial function and these functions follow the exponential function that is robust to noise. Clustering medical images using kernel-based FCM and IFCM clustering is discussed in this chapter. Clustering of tumours/clots of CT scan images of the brain is also carried out. Intuitionistic fuzzy kernel c means clustering performs better on noisy images.

References

1. Ahmed, M.N. et al., A modified fuzzy c-means algorithm for bias field estimation and segmentation of MRI data, *IEEE Transactions on Medical Imaging*, 21, 193–199, 2002.
2. Bezdek, J.C., Hall, L.O., and Clark, L.P., Review of MR segmentation technique in pattern recognition, *Medical Physics*, 10(20), 33–48, 1993.
3. Chaira, T., Ray, A.K., and Salvetti, O., Intuitionistic fuzzy c means clustering in medical image segmentation, in *Proc. of ICAPR 2007*, ISI, Calcutta, India, 2007.
4. Chaira, T., A novel intuitionistic fuzzy C means clustering algorithm and its application to medical images, *Applied Soft Computing*, 11(2), 1711–1717, 2011.
5. Chaira, T., Intuitionistic fuzzy color clustering of human cell images on different color models, *Journal of Intelligent and Fuzzy Systems*, 23, 43–51, 2012.
6. Chaira, T., A novel intuitionistic fuzzy c means color clustering on human cell images, in *Proc. of IEEE World Congress on Nature and Biologically Inspired Computing NaBic-09*, Coimbatore, India, pp. 736–741, December 2009.
7. Chaira, T. and Panwar, A., An Atanassov's intuitionistic fuzzy kernel based clustering for medical image segmentation, *Journal of Computation and Intelligent Systems*, 7(2), 1–11, 2013.
8. Chen, S.C. and Zhang, D.Q., Robust image segmentation using FCM with spatial constrains based on new kernel-induced distance measure, *IEEE Transactions on Systems, Man and Cybernetics*, 34, 1907–1916, 2004.
9. Iakovidis, D.K. et al., Intuitionistic fuzzy clustering with applications in computer vision, *Lecture Notes in Computer Science*, 5259, 764–774, 2008.
10. Jain, A.K. and Dubes, R.C., *Algorithms for Clustering Data*, Prentice Hall, Inc. Upper Saddle River, New Jersey, 1998.
11. Kannan, S.R. et al., Effective fuzzy c-means based kernel function in segmenting medical images, *Computers in Biology and Medicine*, 40, 572–579, 2010.
12. Neiman, H., *Pattern Analysis and Understanding*, 2nd edn., Springer, Berlin, Germany, p. 22, 1990.
13. Rhee, F.C.H. and Hwang, C., A type-2 fuzzy c means clustering algorithm, *Proc. of Joint 9th IFSA World Congress and 20th NAFIPS*, Vancouver, British Columbia, Canada, Vol. 4, pp. 1926–1929, 2001.
14. Yasnoff, W.A. et al., Error measures for scene segmentation, *Pattern Recognition*, 9, 217–231, 1977.
15. Zhang, D.-Q. and Chen, S.-C., A novel kernelized fuzzy C-means algorithm with application in medical image segmentation, *Artificial Intelligence in Medicine*, 32, 37–50, 2004.
16. http://polconsultant.com/conteduc/hematology/micro/index.htm.

References

8

Edge Detection

8.1 Introduction

Edge detection is a fundamental low-level image processing. An edge is a property of an individual pixel, which is calculated from the functional behaviour of an image in its neighbouring pixel. Edge serves in simplifying the analysis of images by drastically reducing the data to be processed and preserving the useful structural information about object boundaries. Edges in an image are the areas with strong intensity contrasts – a jump in intensity from one pixel to the next pixel. There are many ways to perform edge detection, but the majority of methods may be grouped into two categories: (a) gradient and (b) Laplacian. Gradient-based methods detect the edges by looking at the minimum and maximum of the first derivative of the image. Laplacian method searches for the zero crossings in the second derivative of the image in order to find edges. A number of edge detection techniques may be found, but there is not a single method that can detect the edges efficiently. Traditional edge detection methods such as Robert, Sobel and Prewitt operator and Laplacian of Gaussian operator are widely used. Canny proposed an optimal operator for edge detection. Marr and Hildreth's [17] method is based on the zero crossing of the Laplacian operator, which is applied to the Gaussian smoothed image. But in the detection of the zero crossings in the second derivative, the maxima of the gradient are also captured and give false edges. Most of the existing techniques either are very sensitive to noise or do not give satisfactory results in low-contrast areas.

The problem in general edge detectors is that they behave very poorly on the medical image. The quality of edge detection is highly dependent on lighting conditions, density of edges in the scene and noise. Each of them can be handled by adjusting certain values of the edge detector to find threshold values, but these become very tedious when dealing with medical images. The edge of the medical image contains rich information about the image boundary and background. This information is widely used in locating important tissues, the study of anatomical features, etc. Also, an image may contain noise, so whether a pixel is an edge pixel or noise pixel, it depends on the grey value of the pixel and its surrounding pixels. So, smoothing is required in order to remove the noise present in the image, and Gaussian smoothing is the most common filter used.

Fuzzy methods alleviate these problems as they consider the image to be vague and are better suited for edge information detection and noise filtering than the traditional methods. Edge detection using fuzzy logic provides an alternative approach to detect edges. There are various fuzzy and crisp edge detection methods in the literature (see, e.g. [2,5,10,12–15,19]). Fuzzy methods consider the image to be fuzzy as the edges are not clearly defined. In medical images where the images have poor contrast and the edges/boundaries are not properly visible, edge detection becomes very crucial. Edges may be detected using a fuzzy edge detector (FEDGE) that uses some fuzzy templates or fuzzy reasoning that uses linguistic variables. Apart from fuzzy methods, edges may be detected using the intuitionistic fuzzy method. The intuitionistic fuzzy method considers a more number of uncertainties than the fuzzy method, and thus, it seems to be better suited to those types of images where the presence of uncertainty is high (e.g. medical images, remotely sensed images).

There are various ways one can detect the edges, and these are (a) the thresholding-based method, (b) Hough transform method and (c) boundary-based method.

8.1.1 Thresholding Method

When an edge strength is computed using any of the methods such as Canny and Prewitt, thresholding is required to know the existence of edges. Threshold selection is crucial. If the threshold is low, many unwanted edges and irrelevant features are detected, and if the threshold is high, the image will have many missed or fragmented edges. If the edge thresholding is applied to the gradient magnitude image, the resulting edges will be thick and so some type of edge-thinning post-processing is required. A good edge detection method uses a smoothing parameter to remove any noise and, depending on the type of image smoothing parameter, is adjusted to minimize the unwanted noise, and the thresholding parameter is adjusted so that well-formed edges are produced.

A commonly used approach to compute the appropriate thresholds is *thresholding* using *hysteresis* that is used in Canny's edge detector [3]. It uses initially a smoothing parameter (σ) to remove the noise. Gradient image is computed, and then non-maximal suppression is used, which is an edge-thinning technique. A search is carried out to see if the gradient magnitude is the local maxima in the gradient direction. It keeps only those pixels on the edges which have high magnitude. Finally, hysteresis thresholding is done where unwanted edges are removed. If a simple threshold is used, many important edges are removed. In hysteresis thresholding, two thresholds are used. Initially, to find the start of an edge, an upper threshold is used. Then, following the image pixel by pixel, a path is traced whenever the edge strength is above the lower threshold and stops when the value falls below the lower threshold. In this method, the edges are linked and continuous. But still a problem lies in selecting the threshold parameter.

8.1.2 Hough Transform Method

This method is a robust feature extraction method developed by Hough [18] to find the lines in the image. Later, it is extended to detect objects of varying shapes such as ellipse and circles. This method uses an array called an accumulator to detect the existence of a line $y = mx + b$. In Hough transform, the line characteristics are described in terms of slope m and intercept b. The straight line is represented as a point (b, m) in the parameter space. Hough transform uses an accumulator, and the dimension of the accumulator is the number of unknown parameters of the line (here it is 2). For each pixel and its neighbourhood, it determines if the straight line exists at that pixel. If it exists, it will calculate the parameter of that line and then look for the accumulator's bin that the parameters fall into and increase the value of that bin. By looking at the local maxima in the accumulator space, the most likely lines are extracted. It can also be used for curved shapes such as a circle and an ellipse that contain three parameters, making Hough transform 3D. Later, generalized Hough transform is proposed that deals with complicated shapes.

8.1.3 Boundary-Based Method

It is another method for boundary detection introduced by Kass et al. [12]. He developed a novel technique called snake. Snake is an active contour model which is 'an energy minimizing spline guided by external constraint forces and influenced by image forces that pull it toward features such as lines and edges'. Snakes or active contours are computer-generated curves that find the boundary in an image. This method deals with internal (snake) and external (image) forces. Internal forces prevent stretching and bending, and external forces guide towards object boundaries. Traditional snake is initially close to the boundaries and cannot go into the concavities. Later, it is modified such that it can be initialized anywhere in the boundary and can go into the concavities.

As this book is related to the use of intuitionistic fuzzy/Type II fuzzy set, an overview of fuzzy edge detection is given so that the readers will have an idea on fuzzy methods. Fuzzy methods provide better results, but in some images, these methods fail to produce better edge images. In that case, intuitionistic fuzzy set or Type II fuzzy set theories are used as these sets consider more or different types of uncertainties.

8.2 Fuzzy Methods

Fuzzy edge detection is an approach to edge detection that considers the image to be fuzzy. In most of the images where edges are not clearly defined, edges are broken, vague or blurred, making edge detection very difficult. Especially in medical images where the images are poorly contrasted, edge detection becomes very crucial because improper selection of edges may lead to incorrect diagnosis of diseases. In those cases, fuzzy set theory is very useful, which considers vagueness in the edges. It takes into account the vagueness and ambiguity present in an image in the form of a membership function, and then the edges are detected. Edges may be detected using FEDGEs that use some fuzzy templates or fuzzy reasoning that uses some linguistic variables. There are various fuzzy edge detection methods suggested by many authors; here, we shall discuss several fuzzy edge detection methods.

In this section, different types of fuzzy methods, suggested by different authors, are explained in brief:

1. *Fuzzy Sobel edge detector*: This method is suggested by Khamy et al. [15]. Sobel method uses two 3 × 3 masks that calculate the gradient in two directions, which are convolved with the image. Edges are detected using a threshold defined by the user. In the fuzzy Sobel method, the image is divided into two regions: high- and low-gradient regions. If the pixels are having a high difference in the grey level with respect to the neighbourhood, then the pixels are in the fuzzy edge region, and if the pixels are having a low difference in the grey level with respect to the neighbourhood, then the pixels are in the fuzzy smooth region. The boundaries of the two regions are determined from the four threshold values in the difference histogram. The difference histogram is formed by finding the maximum difference in the grey value of each pixel in all eight directions. The gradient of the input image is initially obtained using the Sobel operator. The final edge image is obtained after using simple fuzzy rules that consider the gradient of the image obtained from the Sobel operator.

2. *Entropy-based fuzzy method*: This method was introduced by Khamy et al. [14]. In this method, entropy is used to find the threshold of a gradient image. Initially, a gradient image is formed using the Sobel

operator and then normalized. Then probability partition of the normalized gradient image is then computed using probabilistic distribution in order to partition the image into two regions: edge and smooth regions. The two fuzzy partitions are characterized by the trapezoidal membership function, and the partitions follow a probabilistic distribution. The membership function represents the conditional probability that a pixel is classified into two regions. Then the entropy is used to calculate the threshold. The threshold partitions the image into two regions – edge region and smooth region – which are determined by minimizing the entropy.

3. *Fuzzy median–based edge detector*: Ho and Ohnishi [13] presented FEDGE. It considers several fuzzy edge templates with template values lying between 0 and 1. As it is a fuzzy case, the input image is normalized. Each template is placed on the image pixel, and the fuzzy similarity measure between each template and the subimage (the image area where the template is placed) is calculated to find the existence of an edge at that pixel point. This is done for all the templates. Then using a simple max–min operator, the final edge image is obtained.

4. *Fuzzy divergence–based edge detector*: This method is also a template-based method suggested by Chaira [4]. A set of 16 templates are used that denote the possible direction of edges in an image. The centre of each template is placed at each pixel position over a normalized image. It involves the calculation of fuzzy divergence between the image window and each set of 16 fuzzy templates. The fuzzy divergence measure at each pixel position in the image, where the template was centred, is calculated between each of the elements of the subimage and the template. This is repeated for all 16 fuzzy templates, and using a simple max operator, the edge image is obtained.

8.3 Intuitionistic Fuzzy Edge Detection Method

Edge detection using intuitionistic fuzzy set theory is described in detail in this section. In this method, the membership and non-membership degrees are considered in a fuzzy image. That means it considers more uncertainties. Edge detection in medical images is considered to perform well as medical images contain unclear regions/boundaries. Few intuitionistic fuzzy edge detection techniques on medical images are provided in this chapter.

8.3.1 Template-Based Edge Detection

This method is an extension of fuzzy divergence–based edge detector [4]. For edge detection, a set of 16 fuzzy templates, each of size 3 × 3, representing the edge profiles of different types, are used:

$$
\begin{bmatrix} 0 & b & a \\ 0 & b & a \\ 0 & b & a \end{bmatrix}
\begin{bmatrix} a & a & a \\ 0 & 0 & 0 \\ b & b & b \end{bmatrix}
\begin{bmatrix} a & a & b \\ a & b & 0 \\ b & 0 & 0 \end{bmatrix}
\begin{bmatrix} b & b & b \\ 0 & 0 & 0 \\ a & a & a \end{bmatrix}
$$

$$
\begin{bmatrix} b & a & a \\ 0 & b & a \\ 0 & 0 & b \end{bmatrix}
\begin{bmatrix} b & a & 0 \\ b & a & 0 \\ b & a & 0 \end{bmatrix}
\begin{bmatrix} a & 0 & b \\ a & 0 & b \\ a & 0 & b \end{bmatrix}
\begin{bmatrix} 0 & 0 & 0 \\ b & b & b \\ a & a & a \end{bmatrix}
$$

$$
\begin{bmatrix} a & a & a \\ b & b & b \\ 0 & 0 & 0 \end{bmatrix}
\begin{bmatrix} a & b & 0 \\ a & b & 0 \\ a & b & 0 \end{bmatrix}
\begin{bmatrix} 0 & 0 & 0 \\ a & a & a \\ b & b & b \end{bmatrix}
\begin{bmatrix} 0 & a & b \\ 0 & a & b \\ 0 & a & b \end{bmatrix}
$$

$$
\begin{bmatrix} b & b & b \\ a & a & a \\ 0 & 0 & 0 \end{bmatrix}
\begin{bmatrix} b & 0 & a \\ b & 0 & a \\ b & 0 & a \end{bmatrix}
\begin{bmatrix} b & 0 & 0 \\ a & b & 0 \\ a & a & b \end{bmatrix}
\begin{bmatrix} 0 & 0 & b \\ 0 & b & a \\ b & a & a \end{bmatrix}
$$

A set of sixteen 3 × 3 templates.

The choice of templates is crucial, which reflects the type and direction of edges. The templates are the examples of the edges, which are also the images. a, b and 0 represent the pixels of the edge templates, where the values of a and b are chosen by the trial-and-error method. The best combination is $a = 0.3$, $b = 0.8$, and with these values, the edge-detected results are better. Sixteen templates are considered to be optimum for edge detection.

The size of the templates is less than the size of the image. The centre of each template is placed at each pixel position (i, j) over a normalized image. The intuitionistic fuzzy divergence (IFD) measure at each pixel position (i, j) in the image, where the template was centred, $IFD(i, j)$, is computed between the image window (same size as that of the template) and the template using the max–min relationship, as given in the following equation [5]:

$$
IFD(i, j) = \max_{N}[\min_{r}(IFD(A, B)] \tag{8.1}
$$

'$r = 9$' is the number of elements in the square template ($3^2 = 9$)
N is the total number of templates

The *IFD* between *A* and *B*, *IFD(A, B)*, is computed between each of the elements a_{ij} and b_{ij} of image window *A* and that of template *B*. It is given as

$$IFD(a_{ij},b_{ij}) = \left(2-[1-\mu_A(a_{ij})+\mu_B(b_{ij})]\cdot e^{\mu_A(a_{ij})-\mu_B(b_{ij})} - [1-\mu_B(b_{ij})+\mu_A(a_{ij})]\cdot e^{\mu_B(b_{ij})-\mu_A(a_{ij})}\right)$$

$$+\left(2-[1-(\mu_A(a_{ij})-(\mu_B(b_{ij}))+(\pi_B(b_{ij})-\pi_A(a_{ij}))]\cdot e^{\mu_A(a_{ij})-\mu_B(b_{ij})-(\pi_B(b_{ij})-\pi_A(a_{ij}))}\right.$$

$$\left.-[1-(\pi_B(b_{ij})-\pi_A(a_{ij}))+(\mu_A(a_{ij})-\mu_B(b_{ij}))]\cdot e^{\pi_B(b_{ij})-\pi_A(a_{ij})-(\mu_A(a_{ij})-\mu_B(b_{ij}))}\right)$$

$$(8.2)$$

Normalized values of the (*i*, *j*)th pixel of image *A* are the membership degrees, $\mu_A a_{ij}$, while the values of the template are the membership degrees of the template pixels, $\mu_B b_{ij}$:

$$\pi_A(a_{ij},b_{ij}) = c*(1-\mu_A(a_{ij},b_{ij}))$$

where *c* is a hesitation degree. The value of *c* should be such that $\pi_A(a_{ij}) + \mu_A(a_{ij}) + \nu_A(a_{ij}) = 1$ holds.

IFD(a_{ij}, b_{ij}) is the *IFD* between each element of the template (b_{ij}) and the image window (a_{ij}). It is calculated for all pixel positions of the image. Finally, an *IFD* matrix, which is of the same size as that of the image, is formed. This *IFD* matrix is thresholded and thinned to get an edge-detected image. Then the threshold is selected manually for getting the final edge-detected result.

Example 8.1

An example in Figures 8.1 and 8.2 will illustrate the effectiveness of intuitionistic fuzzy edge detection directly on medical images. Figure 8.1 is a lung image and Figure 8.2 is a brain image.

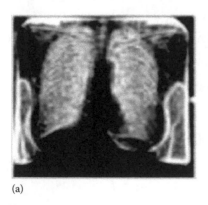 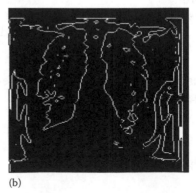

(a)   (b)

FIGURE 8.1
(a) Lung image and (b) edge-detected image. (Modified from Chaira, T. and Ray, A.K., *Appl. Soft Comput.*, 8(2), 919, 2007.)

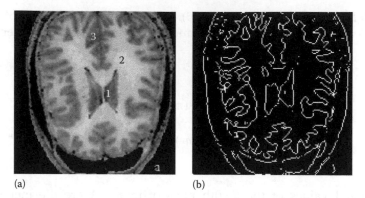

(a)   (b)

FIGURE 8.2
(a) Brain image and (b) edge-detected image. (Modified from Chaira, T. and Ray, A.K., *Appl. Soft Comput.*, 8(2), 919, 2007.)

8.3.2 Edge Detection Using the Median Filter

In this algorithm, the median filter is used to detect the edges of medical images [6]. As medical images are not properly illuminated, direct edge detection techniques sometimes do not work. So, to highlight the edges of the image, edges are enhanced. In blood cell/vessel or nuclei images, nuclei/cells/vessels are not clearly visible, making this technique beneficial. After enhancing the image, edge detection techniques are applied.

An image (say A) of size $M \times N$ is initially fuzzified using the following formula:

$$\mu_F(g) = \frac{g - g_{\min}}{g_{\max} - g_{\min}} \tag{8.3}$$

where
   g is the grey level that ranges from 0 to $L - 1$
   $g_{\min}$ and $g_{\max}$ are the minimum and maximum values of the grey levels of the image, respectively

Based on the fuzzy set, the membership degree of the intuitionistic fuzzy image is calculated as [20]

$$\mu_{IFS}(g;\lambda) = 1 - (1 - \mu_F(g;\lambda))^{\lambda} \tag{8.4}$$

and the non-membership degree is

$$v_{IFS}(g;\lambda) = (1 - \mu_F(g;\lambda))^{\lambda(\lambda+1)}, \quad \lambda \geq 0$$

To obtain an optimum value of λ, the intuitionistic fuzzy entropy (*IFE*) is used.

After the introduction of the non-probabilistic entropy and fuzzy entropy, different types of entropies are given by many authors using intuitionistic fuzzy set theory. In this method, the *IFE* as given by Vlachos and Sergiadis [20] is used and is written as

$$IFE(A;\lambda) = \frac{1}{N \times M} \sum_{i=1}^{N-1} \sum_{j=1}^{M-1} \frac{2\mu_A(g_{ij})v_A(g_{ij}) + \pi_A^2(g_{ij})}{\pi_A^2(g_{ij}) + \mu_A^2(g_{ij}) + v_A^2(g_{ij})} \qquad (8.5)$$

IFE is calculated for all the λ values. The optimum value of λ that corresponds to the maximum value of entropy is computed as

$$\lambda_{opt} = \max(IFE(A;\lambda))$$

With the λ value known, the intuitionistic fuzzy membership degrees of the pixels are computed and an intuitionistic fuzzy image is formed.

Then the median filter of size 3×3 is applied over the intuitionistic fuzzy image. The corresponding area of the image covered by the median filter is the image window. The median value, surrounding the current pixel $A(m, n)$ of the image window, is computed as

$$Z(m,n) = \text{median}\{A(m-i,n-j)\}, \quad (i,j) \in \phi(m,n)$$

where
  $\phi(m, n)$ is the window
  $m = 1, 2, ..., M; \quad n = 1, 2, ..., N$

Total variation in the 3×3 image window with respect to the median of the window is computed as

$$I(m,n) = \sum_{i,j \in W} abs(A(i,j) - Z(m,n)) \qquad (8.6)$$

where
  i and j are the pixels in the 3×3 window
  W is the window

The size of the median filter is the same as the size of the image window. For each pixel position, $I(m, n)$ is computed and a new difference matrix is formed. The new matrix is an edge-enhanced image.

If the difference image contains unwanted lines, the image is filtered using a Gaussian filter. After filtering, the image is then edge detected using standard Canny's edge detector.

Example 8.2

An example in Figures 8.3 and 8.4 illustrates the effectiveness of intuitionistic fuzzy edge detection where the image is initially enhanced to highlight the structures of the images. Figure 8.3 is a live cell image, and Figure 8.4 is the nuclei image. It may be noted that the cells or the nuclei are not clearly visible in the original image. On enhancing the images, the edges are enhanced and then edge detection is carried out.

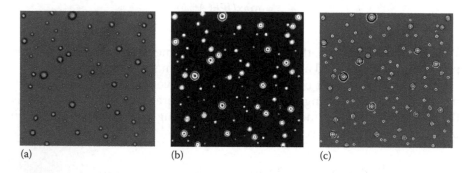

(a)   (b)   (c)

FIGURE 8.3
(a) Nuclei image, (b) edge-enhanced image using the intuitionistic fuzzy method and (c) edge-detected image using the intuitionistic fuzzy method. (Modified from Chaira, T., *Appl. Soft Comput.*, 12(4), 1259, 2012.)

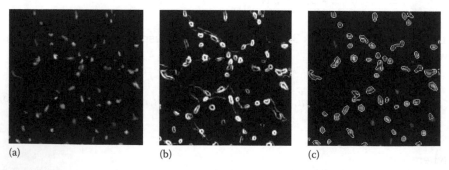

(a)   (b)   (c)

FIGURE 8.4
(a) Nuclei image, (b) edge-enhanced image using the intuitionistic fuzzy method and (c) edge-detected image using the intuitionistic fuzzy method. (Modified from Chaira, T., *Appl. Soft Comput.*, 12(4), 1259, 2012.)

8.4 Fuzzy Edge Image Using Interval-Valued Fuzzy Relation

The fuzzy edge image visually captures the intensity changes, and the image can be considered as an image that represents edges in a fuzzy way. This fact will enable us to better adjust applications where we want to use an edge detector based on fuzzy edge images. Barrenchea et al. [1] had given an idea for the generation of fuzzy edges using the interval-valued fuzzy relation. In the interval-valued fuzzy set, the membership function lies in an interval – upper bound and lower bound.

Interval-valued fuzzy set: Let $L[(0, 1)]$ denote the set of all subintervals of the unit interval $[0, 1]$, that is, $L[(0, 1)] = \{[x_l, x^u] | (x_l, x^u) \in (0, 1)^2, x_l \leq x^u\}$, and x_l and x^u are the lower and upper levels of the interval-valued fuzzy set [1,11], respectively.

Then $L[(0, 1)]$ is a partially ordered set with respect to relation $\leq_L$, which is defined as

$$(x_l, x^u) \leq_L (y_l, y^u) \text{ if and only if } x_l \leq y_l \text{ and } x^u \leq y^u \text{ and } [x_l, x^u], [y_l, y^u] \in L([0, 1])$$

$(L[(0, 1)], \leq_L)$ is a complete lattice with the smallest element $0_L = [0, 0]$ and the largest element $1_L = [1, 1]$.

If U is a universe, the interval-valued fuzzy relation is characterized by mapping $M: U \rightarrow L[(0, 1)]$.

The membership of each element u_i is given as $M(u_i) = [M_l(u_i), M^u(u_i)]$, where M^u and M_l are the upper and lower bounds of the membership range, respectively.

The length of the interval is the difference between the upper and lower bounds.

To construct an interval-valued fuzzy relation, an interval range is required. To generate the lower bound of the interval range, the lower constructor is used, and likewise, the upper constructor is built from the upper bound of the interval range.

The lower constructor is built using *t*-norm and upper constructor using *t*-conorm.

A *t*-norm, $T: [0, 1]^2 \rightarrow [0, 1]$, is an increasing function such that $T(1, x) = x$ for all $x \in [0, 1]$. The three basic *t*-norms are as follows:

1. The minimum *t*-norm by Zadeh, $T_M(x, y) = \min(x, y)$
2. The product *t*-norm by Bandler and Kohout, $T_P(x, y) = x \cdot y$
3. Lukasiewicz *t*-norm, $T_L(x, y) = \max(x + y - 1, 0)$

The associativity extends each *t*-norm to an *n*-ary operation by induction and for each *n*-tuple $\{x_1, x_2, ..., x_n\} \in [0,1]^n$ as in the following:

$$\mathop{T}_{i=1}^{n} x_i = T\left(\mathop{T}_{i=1}^{n-1} x_i, x_n\right) = T(x_1, x_2, x_3, ..., x_n)$$

t-Norm, *T* can take different numbers of arguments.

If $R \in \Re(X \times Y)$ is a relation, T_1 and T_2 are two *t*-norms, and n and m are two values with $n \le (P - 1)/2$ and $n \le (Q - 1)/2$, then the lower constructor using T_1 and T_2 may be defined as

$$L_{T_1,T_2}[R](x,y) = \overset{m}{\underset{\substack{i=-n \\ j=-m}}{\overset{n}{T_1}}} (T_2(R(x-i)(y-j), R(x,y))), \quad \forall (x,y) \in X \times Y \qquad (8.7)$$

where the values of i and j are such that $0 \le x - i \le P - 1$ and $0 \le y - i \le Q - 1$. The values of n and m indicate that the window size is $(2n + 1) \times (2m + 1)$ which is centred at (x, y). An example on how the lower constructor operation works is shown. Considering $n = m = 1$ with window size 3×3, $T_1 = T_2 = T_M$ (min *t*-norm by Zadeh), the value at (x_1, y_1) is computed as

(0, 0)	(0, 1)
0.67	0.77
(1, 0)	(1, 1)
0.74	0.78

$(x-1, y-1)$	$(x-1, y)$	$(x-1, y+1)$
0.76	0.56	0.51
$(x, y-1)$	(x, y)	$(x, y+1)$
0.78	0.81	0.88
$(x+1, y-1)$	$(x+1, y)$	$(x+1, y+1)$
0.62	0.56	0.54

$$L^1_{T_M,T_M}[R](x,y) = \min(\min(0.76, 0.81), (0.56, 0.81), (0.51, 0.81), (0.78, 0.81),$$

$$(0.81, 0.81), (0.88, 0.81), (0.62, 0.81), (0.56, 0.81), (0.54, 0.81))$$

$$= 0.51$$

In this way, the values at each coordinate are computed by shifting the window point by point and a matrix is formed. One may use different *t*-norms for T_1 and T_2. It is to be noted that on applying the lower constructor, the value of R is reduced. The lower constructor darkens the image as it takes lower values (shown in the example). The smaller the *t*-norm, the more reduction in the intensity of the pixels, thereby darkening the image.

Likewise, for the upper constructor, to-conorm is used.

A *t*-conorm, $S : [0, 1]^2 \to [0, 1]$, is an increasing function such that $S(0, x) = x$ for all $x \in [0, 1]$. The three basic *t*-conorms are as follows:

1. The maximum *t*-conorm by Zadeh, $S_M(x, y) = \max(x, y)$
2. The product *t*-conorm by Bandler and Kohout, $S_P(x, y) = x + y - x \cdot y$
3. Lukasiewicz *t*-conorm, $S_L(x, y) = \min(x + y, 1)$

For two norms S_1 and S_2, the upper bound of the level is defined as

$$U_{S_1,S_2}[R](x,y) = \overset{m}{\underset{\substack{i=-n \\ j=-m}}{\overset{n}{S_1}}} (S_2(R(x-i)(y-j), R(x,y)))$$

Following the same procedure as the lower constructor, a new matrix is formed. The upper constructor brightens the image. Max *t*-norm is the smallest *t*-conorm. If the *t*-conorm is large, the intensity of the pixel is high and the image is brighter.

A new fuzzy relation is constructed by taking the difference between the lower and upper constructors:

$$W[R](x,y) = U_{S_1,S_2}[R](x,y) - L_{T_1,T_2}[R](x,y) \tag{8.8}$$

$W[R](x, y)$ is the fuzzy edge that denotes the intensity variation in its neighbourhood. So, in the construction of a fuzzy edge, the length of the interval represents the membership degree of each element in the new fuzzy relation. Fuzzy edge is not the edge image as in Canny; rather, it is the change in the intensity.

8.5 Construction of Enhanced Fuzzy Edge Using Type II Fuzzy Set

The fuzzy edge image can also be obtained using Type II fuzzy set where the membership function in an ordinary fuzzy set is considered as fuzzy [7]. The image is initially normalized to obtain the values in the range [0, 1]. For each pixel, a 3 × 3 neighbourhood is selected and minimum and maximum values are noted. This is done for all the pixels. This way, two image matrices are obtained with maximum and minimum values of the pixel in a 3 × 3 window. As the image is itself fuzzy, the maximum and minimum values are also fuzzy, so for each image matrices, Type II levels are computed.

The upper and lower membership functions for the maximum value image are written as

$$\mu_{max}^{upper} = [\mu_{max}(x)]^{0.75}$$

$$\mu_{max}^{lower} = [\mu_{max}(x)]^{1/0.75}$$

Likewise, the upper and lower membership functions of the minimum valued image matrix are computed.

Fuzzy divergence between the upper and lower levels of the maximum value is computed as

$$Div\left(\mu_{max}^{upper}, \mu_{max}^{lower}\right) = 2 - \left(1 - \mu_{max}^{upper} + \mu_{max}^{lower}\right) \cdot e^{\mu_{max}^{upper} - \mu_{max}^{lower}}$$

$$- \left(1 - \mu_{max}^{lower} + \mu_{max}^{upper}\right) \cdot e^{\mu_{max}^{lower} - \mu_{max}^{upper}} \tag{8.9}$$

Likewise, the divergence between the upper and lower levels of the minimum value is computed as

$$Div\left(\mu_{min}^{upper}, \mu_{min}^{lower}\right) = 2 - \left(1 - \mu_{min}^{upper} + \mu_{min}^{lower}\right) \cdot e^{\mu_{min}^{upper} - \mu_{min}^{lower}}$$

$$- \left(1 - \mu_{min}^{lower} + \mu_{min}^{upper}\right) \cdot e^{\mu_{min}^{lower} - \mu_{min}^{upper}} \tag{8.10}$$

Then the difference between the two divergences – min_divergence and max_divergence – is computed. The difference image obtained is the edge image. In doing so, each pixel is associated with different membership degrees corresponding to its interval length (divergence between the maximum and minimum matrices).

Example 8.3

Three examples shown in Figures 8.5 and 8.6 illustrate the effectiveness of the fuzzy edge image using interval-valued fuzzy relation and Type II fuzzy set. Figure 8.5 is an abnormal RBC (red blood cell) image, and Figure 8.6 is the blood cell image. It is observed that the edges are almost clear using Type II fuzzy set.

8.6 Accurate Edge Detection Technique

In this technique, tumour/haemorrhage/mammogram mass or any other abnormal detection is detected [8] where the method initially clusters, thresholds and then edge detects the image.

The reason for carrying out the steps is as follows:

1. Clustering groups the image into several regions, thus reducing the number of grey levels.
2. Thresholding the clustered image further reduces the number of regions keeping only one concerned region.

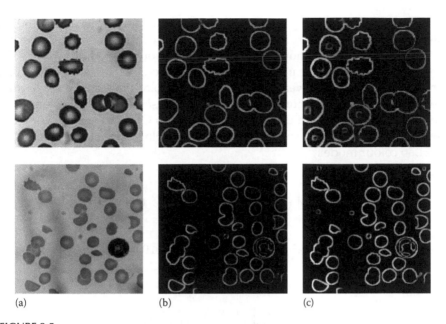

(a)   (b)   (c)

FIGURE 8.5
(a) Abnormal RBC (red blood cell) image, (b) edge image using the interval-valued fuzzy set and (c) edge image using the Type II fuzzy method.

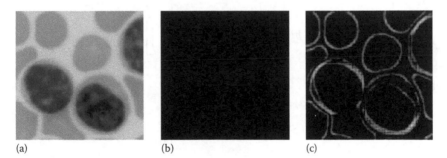

(a)   (b)   (c)

FIGURE 8.6
(a) Blood cell image, (b) edge image using the interval-valued fuzzy set and (c) edge image using Type II fuzzy method.

3. Using the edge detection technique, the edge is obtained for the thresholded image. The threshold required to edge detect the image is appropriate as it considers only one region.

In this method, the intuitionistic fuzzy image is constructed from Sugeno's intuitionistic fuzzy generator. The non-membership function is obtained from Sugeno's intuitionistic fuzzy complement which is written as

$$N(\mu(x)) = \frac{(1-\mu(x))}{(1+\lambda\mu(x))}, \quad \lambda > 0$$

If the tumour/haemorrhage/mammogram cyst or any other region is not properly visible or if the tumour region has a low contrast, pseudo-colouring is done to have some knowledge about the tumour region; otherwise, the greyscale image is kept unchanged. For pseudo-colouring, the RGB (red, green, blue) image is converted into the CIELab colour image as it is a human perceptual model. It is a complete colour space describing all the colours that are visible to the human eye. To convert RGB to CIELAB colour space, the RGB colour space is first converted to the XYZ colour model and then to the CIELab colour model:

$$L^* = 116\left(\frac{Y}{Y_N}\right)^{(1/3)} - 16 \quad \text{for} \left(\frac{Y}{Y_N}\right) > 0.008856$$

$$= 903.3\left(\frac{Y}{Y_N}\right) \qquad \text{for} \left(\frac{Y}{Y_N}\right) \le 0.008856$$

$$a^* = 500\left[f\left(\frac{X}{X_N}\right) - f\left(\frac{Y}{Y_N}\right)\right]$$

$$b^* = 200\left[f\left(\frac{Y}{Y_N}\right) - f\left(\frac{Z}{Z_N}\right)\right]$$

where

$$f(t) = t^{(1/3)}, \quad t > 0.008856 = 7.787t + \frac{16}{116}, \quad t < 0.008856$$

where X_n, Y_n and Z_n are the CIE tristimulus values with reference to the white point (D_{65}).

The image is then segmented into several clusters (user defined). It is followed by histogram thresholding to eliminate the clusters/pixels which are not of interest. So, the image contains only the concerned regions. The thresholded image is edge detected, which is the boundary of the tumour.

The clustering process uses intuitionistic fuzzy clustering [9]. For grey image clustering (i.e. without pseudo-colouring), three features are used: pixel value, mean and standard deviation. For pseudo-coloured image clustering, the CIELab colour model is used where six features are utilized – three for L, a^* and b^* values and three for mean values of each L, a^* and b^* pixel. Mean is calculated for 3×3 neighbourhood of a pixel. A clustered image contains user-defined segments/regions.

As the tumour/haemorrhage region is the region of interest, the image is histogram thresholded to eliminate the unwanted pixels or clusters, thereby keeping only the clusters required by the user. Thus, in a clustered thresholded image, regions that are similar to the tumour/haemorrhage pixel are present. The thresholded image is then processed for edge detection. As there are almost no unwanted regions, edges of the wanted region will be present. The edge detection technique on the thresholded image will detect the exact boundary of the tumour/haemorrhage.

Example 8.4

An example in Figures 8.7 and 8.8 shows the edges of the tumour/clot region where the regions are extracted very clearly.

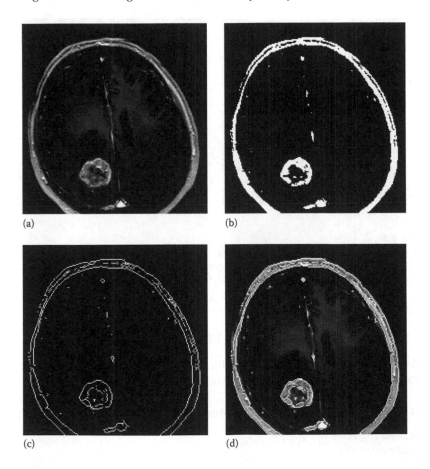

(a)

(b)

(c)

(d)

FIGURE 8.7
(a) Clot image, (b) histogram thresholded clustered image, (c) binary edge image using the intuitionistic fuzzy method and (d) superimposed edge image using the *IFS* method. (Modified from Chaira, T. and Anand, S., *J. Scient. Indust. Res.*, 70(6), 427, 2011.)

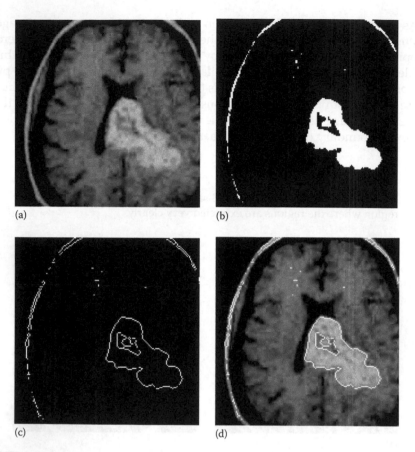

(a)   (b)

(c)   (d)

FIGURE 8.8
(a) Tumour image, (b) histogram thresholded clustered image, (c) binary edge image using the intuitionistic fuzzy method and (d) superimposed edge image using intuitionistic fuzzy method. (Modified from Chaira, T. and Anand, S., *J. Scient. Indust. Res.*, 70(6), 427, 2011.)

8.7 Implementation Using MATLAB®

A MATLAB® code for image edge detection is given, which will be beneficial to the readers to implement the method.

8.7.1 An Example to Find the Edge Image

```
a=imread('group.jpg');
b=rgb2gray(a);
dim=120;
image1=imcrop(b,[1 1 dim-1 dim-1]);
img1=double(image1);
 mn=min(min(img1)); mx=max(max(img1));
```

```
mem1=(img1-mn)./(mx-mn);
 mun=[];
% computing optimum value of 'constant' in intuitionistic fuzzy
  generator
for con=0.1:0.2:10
mem= 1-(1-mem1).^con;
nonmem=(1-mem1).^(con*(con+1));
hes=1-mem-nonmem;
 u=0.0;
   for i=1:dim
     for j=1:dim
       ent= (2*mem(i,j)*nonmem(i,j)+hes(i,j)^2)/(hes(i,j)^2+
       mem(i,j)^2+ nonmem(i,j)^2) + u;
       u=ent;
       end
       end
    lin=ent;
    lin_ind=[mun;lin];
    mun=lin_ind;
    lin_ind;
end
lin_ind
  l=max(lin_ind);
  [con]=find(l==lin_ind);

  fincon= 0.1+(con-1)*0.2   % the constant term
  % computing membership and nonmembership function of an
    intuitionistic fuzzy image
int_mem =1-(1-mem1).^fincon;
nonmem = (1-mem1).^(fincon*(fincon+1));
hes = 1-int_mem-nonmem;
 img = int_mem;
 % median calculation in 3x3 window
 row =dim;col=dim;
 % replicate the edge pixels
edgim=zeros(dim,dim);
dim11=dim+1;dim21=dim+1;
edgimage=zeros(dim11,dim21);
r1=img(2,:);r2=img(row-1,:);c1=img(:,2);c2=img(:,col-1);
  bor1=[0,r1,0];bor2=[0,r2,0];
p1=[c1,img,c2];f2=[bor1;p1;bor2];
f2(1,1)=img(1,1);f2(row+2,col+2)=img(row,col);f2(1,col+2)=img(1,col);
f2(row+2,1)=img(row,1);
 for i=2:dim11
   for j=2:dim21
       im=[f2(i-1,j-1),f2(i,j-1),f2(i+1,j-1);...
       f2(i-1,j),f2(i,j),f2(i+1,j);...
       f2(i-1,j+1),f2(i,j+1),f2(i+1,j+1)];
         cropim1=im(:);
       med_im=median(cropim1);
```

```
         tot_var=sum(sum(abs(im-med_im)));
         diff_im (i-1,j-1)= tot_var;
            end
end
figure,imshow(diff_im)
max(max(diff_im))
fil=fspecial('gaussian',3,0.01);
fil1=uint8(filter2(fil,diff_im));
    edge_im=edge(fil1,'canny',0.45);
    figure,imshow(edge_im);
```

8.7.2 An Example to Find the Fuzzy Edge Image

```
  image=imread('cell2.jpg');
  a1=rgb2gray(image);
  a2=dyaddown(a1,'m',1);a3=dyaddown(a2,'m',1);
  dim1=150;dim2=dim1; row =dim1;col=dim2;
  alpha=0.75;
  img=imcrop(a1,[2 1 dim2-1 dim1-1]);
  figure,imshow(img)
  b1=double(img);
mx=max(max(b1));
img=b1/mx;
% replicate the edge pixels
edgim=zeros(dim1,dim2);
dim11=dim1+1;dim21=dim2+1;
edgimage = zeros(dim11,dim21);
r1=img(2,:); r2=img(row-1,:); c1=img(:,2); c2 = img(:,col-1);
bor1=[0,r1,0];bor2=[0, r2, 0];
p1=[c1,img,c2]; f2=[bor1;p1;bor2];
f2(1,1)=img(1,1); f2(row+2,col+2) =img(row,col);f2(1,col+2) =
  img(1,col);
f2(row+2,1)= img(row,1);
 % calculating the upper and lower membership levels for each window
for i =2:dim11
  for j=2:dim21
    im =
 [f2(i-1,j-1),f2(i,j-1),f2(i+1,j-1);f2(i-1,j),f2(i,j),
   f2(i+1,j);f2(i-1,j+1),f2(i,j+1),f2(i+1,j+1)];
           mxwin=max(max(im)); mnwin=min(min(im));
         % upper and lower membership levels of max value of
           the window
         mxwin_high= mxwin.^alpha;
         mxwin_low=  mxwin.^(1/alpha);
         % upper and lower membership levels of min value of
           the window
         mnwin_high= mnwin.^alpha;
         mnwin_low=  mnwin.^(1/alpha);
```

```
% divergence between the maximum and minimum levels of each
  window
divmx=mxwin_high-mxwin_low;
divmn=mnwin_high-mnwin_low;
  div_mx(i-1,j-1)=divmx;
  div_mn(i-1,j-1)=divmn;
      end
end
  div_im_mx= div_mx/(max(max(div_mx))-min(min(div_mx)));
  div_im_mn= div_mn/(max(max(div_mn))-min(min(div_mn)));
  div_diff=abs(div_im_mx-div_im_mn); % difference in the
    divergence values
figure, imshow(div_diff) % fuzzy edge
```

8.8 Summary

This chapter discusses different edge detection techniques using fuzzy, intuitionistic fuzzy, interval Type II fuzzy and Type II fuzzy set theoretic techniques on medical images. Fuzzy methods consider only one uncertainty and so are useful in medical image processing. But in many cases, fuzzy methods do not show better edge images, and in that case, some advanced fuzzy edge detection techniques are used. Edge enhancement is also shown especially for medical images when edges are not clearly visible, and in that case, edges are enhanced before detection. Fuzzy edge image generation using interval-valued fuzzy set and Type II fuzzy set is also shown. This will help in selecting appropriate edge detectors. A MATLAB code for the methods is also provided which will be beneficial to the readers in implementing the methods.

References

1. Barrenchea, E. et al., Construction of interval valued fuzzy relation with application to generation of fuzzy edge images, *IEEE Transactions on Fuzzy Systems*, 9(5), 819–830, 2011.
2. Becerikli, Y. and Karan, T.M., A new fuzzy approach to edge detection, *Lecture Notes in Computer Science (LNCS)*, 3512, 943–951, June 2005.
3. Canny, J., Computational approach to edge detection, *IEEE Transactions on Pattern Analysis and Machine Intelligence*, 8(6), 679–698, 1986.
4. Chaira, T., Image segmentation and color retrieval: A fuzzy and intuitionistic fuzzy set theoretic approach, PhD Thesis, IIT Kharagpur, India, 2005.

5. Chaira, T. and Ray, A.K., A new measure using intuitionistic fuzzy set theory and its application to edge detection, *Applied Soft Computing*, 8(2), 919–927, 2007.
6. Chaira, T., A rank ordered filter for medical image enhancement and detection using intuitionistic fuzzy set, *Applied Soft Computing*, 12(4), 1259–1266, 2012.
7. Chaira, T. and Ray, A.K., Construction of fuzzy edge image using intuitionistic fuzzy set, *International Journal of Computational Intelligence Systems*, 7(4), 686–695, 2013.
8. Chaira, T. and Anand, S., A novel intuitionistic fuzzy approach for tumour/hemorrhage detection in medical images, *Journal of Scientific and Industrial Research*, 70(6), 427–434, 2011.
9. Chaira, T., Intuitionistic fuzzy color clustering of human cell images on different color models, *International Journal of Intelligent and Fuzzy Systems*, 23, 43–45, 2012.
10. Gómez-Lopera, J. et al., Improved entropic edge detection, in *Proc. of 10th International Conf. of Image Analysis and Processing*, Venice, Italy, pp. 180–184, 1999.
11. Guo-Jun, W. and Ying-Yu, H., Intuitionistic fuzzy sets and L-fuzzy sets, *Fuzzy Sets and Systems*, 110, 271–274, 2000.
12. Ho, K.H.L. and Ohnishi, N., FEDGE – Fuzzy edge detection by fuzzy categorization and classification of edges, *Fuzzy Logic in Artificial Intelligence*, 1188, 182–196, 1995.
13. Kass, M., Wilkin, A., and Terzopaulus, D., Snakes: Active contour models, *International Journal of Computer Vision*, 1, 312–331, 1988.
14. El Khamy, S. et al., Fuzzy edge detection using minimum entropy, in *Proc. of 11th Mediterranean Electrotechnical Conference, MELECON*, 493–508, Cairo, Egypt, 2002.
15. El Khamy, S. et al., Modified Sobel fuzzy edge detector, in *Proc. of IEEE MELECON, 17th National Radio Science Conference*, Minufiya, Egypt, 2000.
16. Li, D. and Cheng, C., New similarity measures of intuitionistic fuzzy sets and application to pattern recognition, *Pattern Recognition Letters*, 23(1–3), 221–225, 2002.
17. Marr, D. and Hildreth, E., Theory of edge detection, *Proceedings of the Royal Society of London: Series B, Biological Sciences*, 207(1167), 187–217, 1980.
18. Sonka, M. et al., *Image Processing, Analysis and Machine Vision*, Brooks/Cole publishing, Pacific Grove, California, United States, 1999.
19. Tao, T.W. and Thomson, W.E., A fuzzy IF-THEN approach to edge detection, in *Proc. of the Second IEEE International Conference on Fuzzy Systems*, pp. 1356–1360, San Francisco, California, 1993.
20. Vlachos, I.K. and Sergiadis, G.D., Intuitionistic fuzzy information – Applications to pattern recognition, *Pattern Recognition Letters*, 28, 197–206, 2007.

9

Fuzzy Mathematical Morphology

9.1 Introduction

Mathematical morphology was born in 1964 from the collaborative work of Georges Matheron and Jean Serra at the École des Mines de Paris, France. It is a theory to analyse and process geometrical structures based on set theory, lattice theory, topology and random functions. It is a tool for extracting different image components that are useful in the representation and description of image regions, boundaries, shapes or skeletons. Initially between the 1960s and 1970s, mathematical morphology dealt with binary images and many binary operators were introduced such as erosion, dilation, opening, closing, skeletonization, hit or miss transform. Later on in the mid-1970s and mid-1980s, it was generalized to greyscale images that require more sophisticated mathematical operations. Simultaneously the operators are extended to new operators. Consequently, mathematical morphology gained much recognition and is used widely in the image processing application.

Mathematical morphology is described almost entirely by set operation such as union, intersection, difference and complement. Set is a collection of pixels in an image. Binary morphology depends only on set membership and does not take into account the grey value or colour of the image pixel.

9.2 Preliminaries on Morphology

As the chapter is related to fuzzy morphology, some basics of morphology are discussed before detailing fuzzy morphology. Dilation and erosion are the basic morphological processing operations. Both dilation and erosion are produced by the interaction of a set called structuring element that has shape and origin and can be described by many ways such as circular, pyramid, linear and square. Dilation is a dual operation of erosion. Dilation dilates the objects and closes holes and gaps of certain shapes and sizes, given by a structuring element. It is an operation that combines two sets using vector addition of a set of elements. For a set of image pixels (x, y) of image $A, f(x, y)$ and structuring element B, dilation is defined as

$$D(A,B) = A \oplus B$$

$$= \{(x + p_x, y + p_y) : (x, y) \in F, (p_x, p_x) \in B\} \quad (9.1)$$

$$= \bigcup_{b \in B}(A + b)$$

Erosion removes the structures of certain shapes and sizes, given by a structuring element, and it shrinks the objects. It is an operation that combines two sets using vector subtraction of a set of elements, or said in another way, erosion of set A by set B is the set intersection of all negative translates of set A by elements of set B or a set of all positions where set B fits inside set A. Erosion of image A by structuring element B is defined as

$$A \ominus B = \{\text{for every } b \in B, \text{ exists an } a \in A \text{ such that } x = a - b\}$$

$$A \ominus B = \{x + b \in A \text{ for every } b \in B\} \quad (9.2)$$

$$\bigcup_{b \in B}(A - b)$$

Again, dilation and erosion can be combined to get an opening or closing operation.

Opening is erosion followed by dilation. Opening can be used to remove small objects, protrusions from objects and connections between objects:

$$A \circ B = (A - B) \oplus B \quad (9.3)$$

Closing is dilation followed by erosion. It removes all holes and gaps in the image objects:

$$A \bullet B = (A \oplus B) \ominus B \quad (9.4)$$

9.2.1 Greyscale Mathematical Morphology

Greyscale morphological operations are an extension of binary morphological operations to greyscale images. The structuring element may be flat where the intensity variation is not continuous or non-flat where the intensity of the structuring element is continuous. If $a(x, y)$ and $b(x, y)$ are the greyscale image and flat structuring element, greyscale dilation is defined as the maximum value of the image in the window outlined by the structuring element b when the origin of b is (x, y):

$$[a \oplus b](x,y) = \max_{(m,n) \in b} \{a(x-m, y-n)\}$$

Greyscale erosion is defined as the minimum value of the image in the window outlined by the structuring element b when the origin of b is (x, y):

$$[a \ominus b](m,n) = \min_{(m,n) \in b} \{a(x+m, y+n)\}$$

The dilated image with a flat structuring element computes the maximum value, so dilation brightens the image. Erosion is opposite to that of dilation.

Dilation and erosion using a non-flat structuring element where the grey values vary over the domain of definition are calculated as follows:

Dilation: $\quad [a \oplus b](x,y) = \max_{(m,n) \in b} \{a(x-m, y-n) + b(x,y)\}$ \hfill (9.5)

Erosion: $\quad [a \ominus b](m,n) = \min_{(m,n) \in b} \{a(x+m, y+n) - b(x,y)\}$ \hfill (9.6)

Greyscale opening and closing are defined as

$$f \circ b = (f \ominus b) \oplus b$$
$$f \cdot b = (f \oplus b) \ominus b$$
\hfill (9.7)

9.3 Fuzzy Mathematical Morphology

In digital image processing, fuzzy set theory has found a promising field of application. Fuzzy mathematical morphology is developed to soften the binary morphology to make the operators less sensitive to image imprecision. It is an alternative to greyscale morphology. It is studied in terms of fuzzy fitting [3,5,8,9,10,12,13,14,19]. The fuzziness is introduced with the degree to which the structuring element fits into the image. The morphological operations are modelled on a fuzzy notion. Fuzziness is introduced only in modelling greyscale images and not in the operations. Fuzzy mathematical morphological operations are obtained by replacing ordinary set theoretic operations with fuzzy set. Fuzzy morphology was first defined by Goecgherian [9]. Since then, many authors extended the framework in their work.

Here, a method to generate adjoint fuzzy morphological operations using a conjunctor and implicator [8,16] is discussed.

Mapping $c(x, y)$: $[0, 1] \times [1, 0] \mapsto [0, 1]$ is a fuzzy conjunction if c is increasing in both x and y. Each conjunctor satisfies

$$c(x, 0) = c(0, x) = 0, \quad x \in [0, 1]$$

Mapping $i(x, y)$: $[0, 1] \times [1, 0] \mapsto [0, 1]$ is a fuzzy implication if i is increasing in y and decreasing in x. Each implicator satisfies

$$i(x, 1) = i(0, x) = 1, \quad x \in [0, 1]$$

A conjunctor is a *t*-norm if it is commutative, that is, $c(x, y) = c(y, x)$, and associative

$$c(c(x, y), z) = c(x, c(y, z)) \quad \text{and} \quad c(x, 1) = x$$

Deng and Heijmans [7] proposed a number of conjunctor–implicator pairs to construct morphological operations. Some examples on conjunctor–implicator are given in the following:

Gödel–Brouwer operation

$$c(b,y) = \min(b,y)$$

$$i(b,x) = \begin{matrix} x, & \text{for } x < b \\ 1, & \text{for } x \geq b \end{matrix} \tag{9.8}$$

Lukasiewicz operation

$$c(b,y) = \max(0, b+y-1)$$
$$i(b,x) = \min(1, x-b+1) \tag{9.9}$$

Hamacher operation for $\lambda > 1$

$$c(b,y) = \begin{matrix} \dfrac{y}{\lambda+(1-\lambda)(b+y-by)}, & \text{for } b > 0 \\ 0, & \text{for } b = 0 \end{matrix}$$

$$i(b,x) = \begin{matrix} \dfrac{(1-\lambda)bx+\lambda x}{\lambda+(1-\lambda)(1-x+bx)}, & \text{for } b > 0 \\ 1, & \text{for } b = 0 \end{matrix} \tag{9.10}$$

If $0 \le \lambda \le 1$, then

$$c(b,y) = \dfrac{(1-\lambda)by + \lambda y}{\lambda + (1-\lambda)(1-y+by)}, \qquad \text{for } b > 0$$
$$\qquad\qquad 0, \qquad\qquad\qquad\qquad\qquad \text{for } b = 0 \tag{9.11}$$

$$i(b,x) = \dfrac{x}{\lambda + (1-\lambda)(b+x-bx)}, \qquad \text{for } b > 0$$
$$\qquad\qquad 1, \qquad\qquad\qquad\qquad\qquad \text{for } b = 0$$

Reichenbach

$$C(b,y) = \begin{matrix} 0, & t \le 1-b \\ t, & t > 1-b \end{matrix} \tag{9.12}$$

$$I(b,s) = 1 - b + bx$$

An adjoint pair of fuzzy erosion and dilation is given as follows [15]:

$$D_B(A)(x) = \sup_{y} c(B(x-y), A(y))$$
$$E_B(A)(x) = \inf_{y} i(B(y-x), A(y)) \tag{9.13}$$

Several definitions for fuzzy erosion and dilation are proposed by many authors.

9.3.1 Different Definitions of Fuzzy Morphology

All the definitions are given for any fuzzy set A and structuring element B, which are defined over a space S at any point 'x'.

1. Definition by De Baets and Kerre [1] and Bloch and Maitre [3]

$$D1A(x) = \sup_{y \in S} \min\left[A(y), B(y-x) \right]$$
$$E1A(x) = \inf_{y \in S} \max\left[A(y), 1 - B(y-x) \right] \tag{9.14}$$

2. Definition by Bloch and Maitre [3]

$$D2A(x) = \sup_{y \in S} \left[A(y), B(y-x) \right]$$
$$E2A(x) = \inf_{y \in S} \left[A(y), 1 - B(y-x) \right] \tag{9.15}$$

3. Definition by Sinha and Dougherty [16,17]

$$D3A(x) = \sup_{y \in S} \max\left[0, A(y) + B(y - x) - 1\right]$$

$$E3A(x) = \inf_{y \in S} \min\left[1, 1 + A(y) - B(y - x)\right]$$

(9.16)

4. Another definition by Sinha and Dougherty [16,17]

$$D4A(x) = \sup_{y \in S} \max\left[0, 1 - \lambda(A(y)) - \lambda(B(y - x))\right]$$

$$E4A(x) = \inf_{y \in S} \min\left[1, \lambda(1 - A(y)) + \lambda(B(y - x))\right]$$

(9.17)

where λ is a function from [0, 1] to [1, 0] and satisfies the following conditions:
   a. $\lambda(x)$ is a decreasing function of x.
   b. $\lambda(0) = 1$ and $\lambda(1) = 0$.
   c. The equation $\lambda(x) = \alpha$ has a single solution, $\forall x \in [0.5, 1]$.
   d. $\lambda(x) = 0$ has a single solution.
   e. $\lambda(x) + \lambda(1 - x) \geq 1, \forall x \in [0, 1]$.

When $\lambda(x) = 1 - x$, $D4\mu(x)$ and $E4\mu(x)$ reduce to $D3\mu(x)$, $E3\mu(x)$.

9.3.2 Fuzzy Morphology Using Lukasiewicz Operator

Lukasiewicz generalized operator maps $L: [0, 1] \times [0, 1] \rightarrow [0, 1]$ such that

$$L(a,b) = \min\left[1, \lambda(a) + \lambda(1 - b)\right], \quad \forall a, b \in [0, 1]$$

(9.18)

where $\lambda: [0, 1] \rightarrow [0, 1]$ and $\lambda(0) = 1$ and $\lambda(1) = 0$.

The properties of Smets–Magrez's axioms [1] for implication operators are used for the properties of Lukasiewicz implication. These are as follows:

1. The value $L(a, b) = L(1 - b, 1 - a), \forall a, b \in 1$.
2. $L(1, b) = b, \forall b \in [0, 1]$.
3. $L(a, b)$ is continuous.
4. $L(\cdot, b)$ is non-increasing $\forall b \in [0, 1]$ and $L(a, \cdot)$ is non-decreasing $\forall a \in [0, 1]$.
5. $a \leq b \Leftrightarrow L(a, b) = 1, \forall a, b \in 1$.
6. The value $L(a, b)$ depends on the value of a and b.

From the earlier properties of Smets–Magrez's axioms, the properties of Lukasiewicz operators are as follows:

1. $L(0, b) = 1, \forall b \in [0, 1]$
2. $L(a, 0) = \lambda(a), \forall a \in [0, 1]$
3. $L(a, 1) = 1, \forall a \in [0, 1]$
4. $L(1, b) = \lambda(1 - b), \forall b \in [0, 1]$
5. $L(a, b) = 1 \Leftrightarrow \lambda(a) + \lambda(1 - b) \geq 1, \forall a, b \in 1$
6. $L(a, b) = 0 \Leftrightarrow \lambda(a)$ and $\lambda(1 - b) = 0$

Lukasiewicz erosion and dilation [5,18] for an image A by another image B (A and B are the fuzzy sets of a subset) using inclusion grade R.

The inclusion grade $R \in F(S) \times F(S)$ between fuzzy subsets in some universe S is

$$R[A, B] = \inf_{x \in S} \left\{ \min\left[1, \lambda[A(x)] + \lambda[1 - B(x)] \right] \right\}, \quad \forall A, B \in F(S)$$

and it satisfies the following conditions:

$$R(A, B) = 1 \Leftrightarrow A \subseteq B$$

$$R(A, B) = 0 \Leftrightarrow \exists x \in U, A(x) = 1, B(x) = 0$$

$$R(A, B) = R(BC, AC)$$

$$R((A \cup B), C) \geq \min(R(A, C), \min(B, C))$$

$$R(A, B \cap C) \geq \min(R(A, B), \min(A, C))$$

Dilation and erosion are defined as

$$E_L(A, B)(x) = R[B(x), A(x)]$$
$$= \inf_{x \in S} \left\{ \min\left[1, \lambda[B(x)] + \lambda[1 - A(x)] \right] \right\} \tag{9.19}$$

$$D_L(A, B)(x) = 1 - R[B(x), A(x)]$$
$$= 1 - \inf_{x \in S} \left\{ \min\left[1, \lambda[B(x)] + \lambda[1 - A(x)] \right] \right\} \tag{9.20}$$

From Equation 9.18, a wide class of operators can be defined with different values of λ. These are

1. $\lambda_n(x) = (1-x)^n, \quad n \geq 1$ (9.21)

This is equivalent to Zadeh's concentration operation on linguistic hedges:

2. $\lambda_n(x) = \dfrac{1-x}{1+((1/n)-1)x}, \quad n \geq 1$ (9.22)

This is equivalent to Sugeno's λ complementation:

3. $\lambda_n(x) = \dfrac{1}{1+x^n} - \dfrac{x}{2}, \quad n \geq \dfrac{\ln(3)}{\ln(2)} = 1.58$ (9.23)

4. $\lambda_n(x) = \dfrac{e^{1-x^n}-1}{e-1}, \quad n \geq \dfrac{\ln(-\ln((e+1)/2e))}{\ln(1/2)} = 1.396$ (9.24)

Equations 9.21 and 9.22 are widely used. With different selections of $\lambda(x)$ and with different values of 'n', different eroded and dilated images are obtained.

9.3.3 Fuzzy Morphology Using *t*-Norms and *t*-Conorms by De Baets and Kerre and Bloch and Maitre

$$D5A(x) = \sup_{y \in S} i\left[B(y-x), A(y)\right]$$

$$E5A(x) = \inf_{y \in S} u\left[c \cdot \left(B(y-x)\right), A(y)\right]$$
                                                       (9.25)

where
    i is any *t*-norm (fuzzy intersection)
    u is any *t*-conorm (fuzzy union)
    c is a complement
    A is an image
    B is a structuring element

Using different *t*-norms (T) and *t*-conorms (T^*), different forms of erosion and dilation can be generated:

    1. Zadeh *t*-operators

$$T^* = \max(x, y)$$

$$T = \min(x, y)$$

From the definition of dilation,

$$DA(x) = \sup_{y \in S} i\left[B(y-x), A(y)\right]$$

Substituting Zadeh's *t*-norm, we get

$$DA(x) = \sup_{y \in S} \min\left[B(y-x), A(y)\right]$$

Likewise, for erosion, $EA(x) = \inf_{y \in S} u\left[c \cdot (B(y-x)), A(y)\right]$
'*c*' is the complement, so $c \cdot (B(y-x)) = 1 - B(y-x)$

So,

$$EA(x) = \inf_{y \in S} \max\left[A(y), B(y-x)\right]$$

2. Bandler and Kohout [2] *t*-operators

$$\text{T-conorms:} \quad T^* = x + y - xy$$

$$\text{T-norm:} \quad T = xy$$

$$\text{Fuzzy erosion:} \quad EA(x) = \inf_{y \in S}\left[A(y) \cdot B(y-x) + 1 - B(y-x)\right]$$

$$\text{Fuzzy dilation:} \quad D(A(x)) = \sup_{y \in S}\left[A(y) \cdot B(y-x)\right]$$

3. Hamacher *t*-operators [11]

$$T = \frac{x \cdot y}{\gamma + (1-\gamma) \cdot (x + y - x \cdot y)}$$

$$T^* = \frac{x + y - x \cdot y - (1-\gamma)xy}{1 - (1-\gamma) \cdot xy}$$

Erosion and dilation are defined as

Fuzzy erosion:

$$EA(x) = \inf_{y \in S}\left[\frac{1 - B(y-x) + A(y) - A(y) \cdot B(y-x) - (1-\gamma) \cdot A(y) \cdot (1 - B(y-x))}{1 - (1-\gamma) \cdot A(y) \cdot (1 - B(y-x))}\right]$$

Fuzzy dilation: $DA(x) = \sup\limits_{y \in S}\left[\dfrac{A(y) \cdot B(y-x)}{\gamma + (1-\gamma) \cdot A(y) + B(y-x) - A(y) \cdot B(y-x)} \right]$

4. Lukasiewicz *t*-operators

$$T = \max(0, x+y-1)$$

$$T^* = \min(1, x-y+1)$$

Fuzzy erosion: $E(A,B)(x) = \inf\limits_{y \in S} \max\left(0, 1+A(y)-B(y-x)\right)$

Fuzzy dilation: $D(A,B)(x) = \sup\limits_{y \in S} \min(1, A(y)+B(y-x)-1)$

9.4 Opening and Closing Operations

Fuzzy opening and closing operations are defined using fuzzy erosion and dilation:

$$O(A,B)(x) = D(E(A,B)(x), B(x))$$

$$= D\left(\inf\limits_{x \in S} \min[1, 1+A(x)-B(x)], B(x) \right)$$

$$C(A,B)(x) = E(D(A,B)(x), B(x))$$

$$= E\left(\sup\limits_{x \in S} \max[0, A(x)+B(x)-1], B(x) \right) \left(\text{using Sinha and Dougherty}\right)$$

Using Lukasiewicz operator, opening and closing of an image are defined as

$$O_L(A,B)(x) = D^L\left[E^L(A,B)(x), B(x) \right]$$

$$= D^L\left[\inf\limits_{x \in S} L[B(x), A(x)], B(x) \right]$$

$$C_L(A,B)(x) = E^L\left[D^L(A,B)(x), B(x) \right]$$

$$= E^L\left[1 - \inf\limits_{x \in S} L[B(x), A(x)], B(x) \right] \quad L \text{ is a Lukasiewicz operator}$$

It is to be noted that the principle of duality plays an important role in fuzzy dilation and erosion, since one operation can be deduced from another. If the complementation is $\mu^C(x) = 1 - \mu(x)$, then the duality with respect to the complementation between erosion and dilation in relation to the structuring element is expressed as [3]

$$D(\mu) = 1 - E(1-\mu)(x)$$

Example: An example of a matrix is shown for dilation and erosion (using Sinha and Dougherty) where erosion and dilation are defined using the membership function. Consider A as a fuzzy image, B as a fuzzy structuring element and μ_A and μ_B as the membership functions of the image and the structuring element, respectively.
Erosion and dilation are defined as

$$E(A,B) = \mu_{A\ominus B}(x) = \min\left[\min_{y\in B}[1+\mu_A(x+y)-\mu_B(y)]\right]$$

$$= \min\left[1, \min_{y\in B}[1+\mu_A(x+y)-\mu_B(y)]\right]$$

$$D(A,B) = \mu_{A\oplus B}(x) = \max\left[\max_{y\in B}[\mu_A(x-y)+\mu_B(y)-1]\right]$$

$$= \max\left[0, \max_{y\in B}[\mu_A(x-y)+\mu_B(y)-1]\right]$$

Fuzzy erosion and dilation have membership functions within the interval [0, 1]:

$$\mu_A = \begin{bmatrix} 0.3 & 1.0 & 0.8 & 0.9 & 0.4 \\ 0.4 & 0.8 & 0.9 & 1.0 & 0.5 \\ 0.4 & 0.9 & 0.3 & 0.8 & 0.3 \end{bmatrix}$$

$$\mu_B = [0.8, 0.9], \quad \mu_{-B} = [0.8, 0.9]$$

The arrows denote the origin of the coordinate system.

Fuzzy erosion is computed as

$$\mu_E(0,0) = \min[1, \min(0.3 - 0.8 + 1, 1 - 0.9 + 1)]$$
$$= \min[1, \min(0.5, 1.1)] = \min(1, 0.5) = 0.5$$

$$\mu_E(0,1) = \min[1, \min(0.3 - 0.7 + 1, 1 - 0.8 + 1, 0.8 - 0.9 + 1)]$$
$$= \min[1, \min(0.6, 1.2, 0.9)] = \min(1, 0.6) = 0.6$$

$$\mu_E(0,2) = \min[1, \min(1 - 0.7 + 1, 0.8 - 0.8 + 1, 0.9 - 0.9 + 1)]$$
$$= \min[1, \min(1.3, 1, 1)] = 1$$

$$\mu_E(0,3) = \min[1, \min(0.8 - 0.7 + 1, 0.9 - 0.8 + 1, 0.4 - 0.9 + 1)]$$
$$= \min[1, \min(1.1, 1.1, 0.5)] = 0.5$$

$$\mu_E(0,4) = \min[1, \min(0.9 - 0.7 + 1, 0.4 - 0.8 + 1)]$$
$$= \min[1, \min(1.2, 0.6)] = 0.6$$

$$\mu_E(1,0) = \min[1, \min(0.4 - 0.8 + 1, 0.8 - 0.9 + 1)]$$
$$= \min[1, \min(0.6, 0.9)] = 0.6$$

$$\mu_E(1,1) = \min[1, \min(0.4 - 0.7 + 1, 0.8 - 0.8 + 1, 0.9 - 0.9 + 1)]$$
$$= \min[1, \min(0.7, 1, 1)] = 0.7$$

$$\mu_E(1,2) = \min[1, \min(0.8 - 0.7 + 1, 0.9 - 0.8 + 1, 1 - 0.9 + 1)]$$
$$= \min[1, \min(1.1, 1.1, 1.1)] = 1$$

$$\mu_E(1,3) = \min[1, \min(0.9 - 0.7 + 1, 1 - 0.8 + 1, 0.5 - 0.9 + 1)]$$
$$= \min[1, \min(1.2, 1.2, 0.6)] = 0.6$$

$$\mu_E(1,4) = \min[1, \min(1 - 0.7 + 1, 0.5 - 0.8 + 1)]$$
$$= \min[1, \min(1.3, 0.7)] = 0.7$$

$$\mu_E(2,0) = \min[1, \min(0.4 - 0.8 + 1, 0.9 - 0.9 + 1)]$$
$$= \min[1, \min(0.6, 1)] = 0.6$$

$$\mu_E(2,1) = \min[1, \min(0.4 - 0.7 + 1, 0.9 - 0.8 + 1, 0.3 - 0.9 + 1)]$$
$$= \min[1, \min(0.7, 1.1, 0.4)] = 0.4$$

$$\mu_E(2,2) = \min[1, \min(0.9 - 0.7 + 1, 0.3 - 0.8 + 1, 0.8 - 0.9 + 1)]$$
$$= \min[1, \min(1.2, 0.5, 0.9)] = 0.5$$

$$\mu_E(2,3) = \min[1, \min(0.3 - 0.7 + 1, 0.8 - 0.8 + 1, 0.3 - 0.9 + 1)]$$
$$= \min[1, \min(0.6, 1, 0.4)] = 0.4$$

$$\mu_E(2,4) = \min[1, \min(0.8 - 0.7 + 1, 0.3 - 0.8 + 1)]$$
$$= \min[1, \min(1.1, 0.5)] = 0.5$$

The eroded image is

$$\begin{bmatrix} 0.5 & 0.6 & 1.0 & 0.5 & 0.6 \\ 0.6 & 0.7 & 1.0 & 0.6 & 0.3 \\ 0.6 & 0.4 & 0.5 & 0.4 & 0.5 \end{bmatrix}$$

Likewise, for dilation,

$$\mu_E(0,0) = \max[0, \max(0.3 + 0.9 - 1)] = \max[0, 0.2] = 0.2$$

$$\mu_E(0,1) = \max[0, \max(0.3 + 0.8 - 1, 1 + 0.9 - 1)]$$
$$= \max[0, \max(0.1, 0.9)] = 0.9$$

$$\mu_E(0,2) = \max[0, \max(0.3 + 0.7 - 1, 1 + 0.8 - 1, 0.8 + 0.9 - 1)]$$
$$= \max[0, \max(0, 0.8, 0.7)] = 0.8$$

$$\mu_E(0,3) = \max[0, \max(1 + 0.7 - 1, 0.8 + 0.8 - 1, 0.9 + 0.9 - 1)]$$
$$= \max[0, \max(0.7, 0.6, 0.8)] = 0.8$$

$$\mu_E(0,4) = \max[0, \max(0.8 + 0.7 - 1, 0.8 + 0.9 - 1, 0.9 + 0.4 - 1)]$$
$$= \max[0, \max(0.5, 0.7, 0.3)] = 0.7$$

$$\mu_E(1,0) = \max[0, \max(0.4 + 0.9 - 1)] = 0.3$$

$$\mu_E(1,1) = \max[0, \max(0.4 + 0.8 - 1, 0.8 + 0.9 - 1)]$$
$$= \max[0, \max(0.2, 0.7)] = 0.7$$

$$\mu_E(1,2) = \max[0, \max(0.4 + 0.7 - 1, 0.8 + 0.8 - 1, 0.9 + 0.9 - 1)]$$
$$= \max[0, \max(0.1, 0.6, 0.8)] = 0.8$$

$$\mu_E(1,3) = \max[0, \max(0.8 + 0.7 - 1, 0.9 + 0.8 - 1, 1 + 0.9 - 1)]$$
$$= \max[0, \max(0.5, 0.7, 0.9)] = 0.9$$

$$\mu_E(1,4) = \max[0, \max(0.7 + 0.9 - 1, 1 + 0.8 - 1, 0.5 + 0.9 - 1)]$$
$$= \max[0, \max(0.6, 0.8, 0.4)] = 0.8$$

$$\mu_E(2,0) = \max[0, \max(0.4 + 0.9 - 1)] = 0.3$$

$$\mu_E(2,1) = \max[0, \max(0.4 + 0.8 - 1, 0.9 + 0.9 - 1)]$$
$$= \max[0, \max(0.2, 0.8)] = 0.8$$

$$\mu_E(2,2) = \max[0, \max(0.4 + 0.7 - 1, 0.9 + 0.8 - 1, 0.3 + 0.9 - 1)]$$
$$= \max[0, \max(0.1, 0.7, 0.2)] = 0.7$$

$$\mu_E(2,3) = \max[0, \max(0.7 + 0.9 - 1, 0.3 + 0.8 - 1, 0.8 + 0.9 - 1)]$$
$$= \max[0, \max(0.6, 0.1, 0.7)] = 0.7$$

$$\mu_E(2,4) = \max[0, \max(0.3 + 0.7 - 1, 0.8 + 0.8 - 1, 0.3 + 0.9 - 1)]$$
$$= \max[0, \max(0.0, 0.6, 0.2)] = 0.6$$

The dilated image is

$$
\begin{bmatrix}
0.2 & 0.9 & 0.8 & 0.8 & 0.7 \\
0.3 & 0.7 & 0.8 & 0.9 & 0.8 \\
0.3 & 0.8 & 0.7 & 0.7 & 0.6
\end{bmatrix}
$$

9.5 Fuzzy Morphology in Image Processing

Morphological operators transform the original image into another image of certain shape and size, also known as structuring element. Mathematical morphology provides an approach to analyse the geometric characteristics of images and has been widely used in image edge detection, segmentation, noise suppression and so on. Fuzzy mathematical morphology extends the binary morphological operators to grey-level images. In binary morphology, fuzzy erosion, dilation, opening and closing are present. In a similar way in fuzzy morphological operations, union operation is replaced by a maximum operation and intersection operation is replaced by a minimum operation.

Morphological operators are used to find the morphological gradient or to denoise the image. The effect of erosion and dilation operations is better for finding the image edge by taking the difference between the dilated image and eroded image, but they do not perform well in noisy images. As opposed to erosion and dilation, opening and closing operations perform better in denoising the images.

Techniques for mathematical morphology using fuzzy set and intuitionistic fuzzy set on medical images are discussed.

9.5.1 Edge Detection

The image is initially fuzzified to have the values between [0, 1]. The structuring element selected is

$$\begin{bmatrix} 0.86 & 0.86 & 0.86 \\ 0.86 & 1 & 0.86 \\ 0.86 & 0.86 & 0.86 \end{bmatrix}$$

From Lukasiewicz's definition, the *t*-norm and *t*-conorm are

$$T(A,B) = \max(0, A+B-1)$$

$$T^*(A,B) = \min(1, A+B)$$

where
  A is an image
  B is a structuring element

From the definition of dilation and erosion,

$$D(A,B) = \sup_{y \in S} i\left[B(y-x), A(y)\right]$$

$$E(A,B) = \inf_{y \in S} u\big[c \cdot B(y-x), A(y)\big]$$

where
   i is a *t*-norm
   u is a *t*-conorm

So,

$$D(A,B) = \sup_{y \in S}(\max(0, A(y) + B(y-x) + 1))$$

$$E(A,B) = \inf_{y \in S}(\min(1, A(y) + c \cdot B(y-x)))$$

$$= \inf_{y \in S}(\min(1, A(y) + 1 - B(y-x)))$$

$$= \inf_{y \in S}(\min(1, 1 + A(y) - B(y-x)))$$

The image is initially fuzzified using any membership function. Then erosion and dilation operations are performed using the structuring element. The difference of the eroded and dilated image gives a gradient image.

9.5.2 Intuitionistic Fuzzy Approach

Another approach to find the edge image is by using Hamacher *t*-norm and *t*-conorm using intuitionistic fuzzy set theory. The image is converted to an intuitionistic fuzzy image, and then using morphological operators, the edge image is obtained.

From Sugeno's fuzzy complement or intuitionistic fuzzy generator, the intuitionistic fuzzy membership function is computed as

$$\mu'_{mn} = 1 - \frac{1-\mu_{mn}}{1+\lambda \cdot \mu_{mn}} = \frac{(1+\lambda) \cdot \mu_{mn}}{1+\lambda \cdot \mu_{mn}}$$

Using Sugeno-type fuzzy negation,

$$\varphi(x) = \frac{1-\mu_{mn}}{1+\lambda \cdot \mu_{mn}}$$

The non-membership function of an intuitionistic fuzzy image is computed as

$$v'_{mn} = \varphi(\mu'_{mn}) = \frac{1-\mu'_{mn}}{1+\lambda \cdot \mu'_{mn}} = \frac{1 - \dfrac{(1+\lambda)\mu_{mn}}{(1+\lambda \cdot \mu_{mn})}}{1 + \dfrac{\lambda \cdot (1+\lambda)\mu_{mn}}{(1+\lambda \cdot \mu_{mn})}}$$

$$= \frac{1-\mu_{mn}}{1+2 \cdot \lambda \cdot \mu_{mn} + \lambda^2 \mu_{mn}}$$

where μ_{mn} is the membership function of the fuzzified image that is calculated as

$$\mu_{mn} = \frac{g_{mn} - g_{min}}{g_{max} - g_{min}}$$

g_{max} and g_{min} are the maximum and minimum grey values of the image, respectively.

λ is calculated using the intuitionistic fuzzy entropy [6]:

$$IE(A) = \frac{1}{N} \sum_{n=1}^{N} \sum_{m=1}^{M} \pi_A(x_{mn}) \cdot e^{1 - \pi_A(x_{mn})}$$

$\pi_{mn}, \pi_{mn} = 1 - \mu'_{mn} - v'_{mn}$, is the hesitation degree.

The optimum value of λ is

$$\lambda_{opt} = \max_{\lambda}(IE(A, \lambda))$$

The structuring element used is

$$\begin{bmatrix} 0.8 & 0.2 & 0.8 \\ 0.2 & 0.8 & 0.2 \\ 0.8 & 0.2 & 0.8 \end{bmatrix}$$

Erosion and dilation are defined as

$$D(A, B) = \sup_{y \in S} i\left[B(y - x), A(y)\right]$$

$$E(A, B) = \inf_{y \in S} u\left[c \cdot B(y - x), A(y)\right]$$

where
  i is any *t*-norm (fuzzy intersection)
  u is any *t*-conorm (fuzzy union)
  c is a complement

Using *t*-norm and *t*-conorm, erosion and dilation are defined as

$$E(A, B) = \inf_{y \in S}\left[\frac{1 - B(y - x) + A(y) - A(y) \cdot B(y - x) - (1 - \gamma)A(y) \cdot (1 - B(y - x))}{1 - (1 - \gamma) \cdot A(y) \cdot (1 - B(y - x))}\right]$$

$$D(A,B) = \sup_{y \in S} \left[\frac{A(y) \cdot B(y-x)}{\gamma + (1-\gamma) \cdot (A(y) + B(y-x) - A(y) \cdot B(y-x))} \right]$$

The value of $\gamma = 0.2$ is chosen by the trial-and-error method. Next, the edge image is created by taking the difference of the dilation and erosion, and then the image is binarized to obtain a binarized edge image.

Example 9.1

Three results are shown on medical images in Figures 9.1 through 9.3. Results on fuzzy morphology using Lukasiewicz operator and intuitionistic fuzzy morphology using Hamacher t-operators are shown.

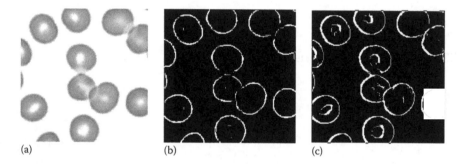

(a) (b) (c)

FIGURE 9.1
(a) Blood cell image, (b) edge image using Lukasiewicz operator and (c) edge image using Hamacher t-operator utilizing the intuitionistic fuzzy approach.

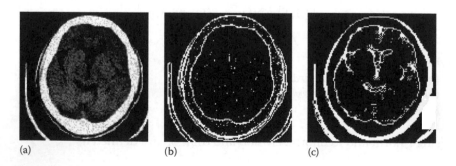

(a) (b) (c)

FIGURE 9.2
(a) Noisy CT scan brain image, (b) edge image using Lukasiewicz operator and (c) edge image using Hamacher t-operator utilizing the intuitionistic fuzzy approach.

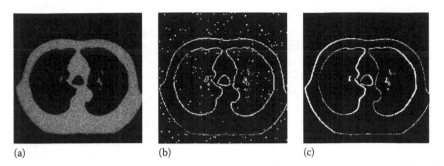

(a)   (b)   (c)

FIGURE 9.3
(a) Noisy heart image, (b) edge image using Lukasiewicz operator and (c) edge image using Hamacher *t*-operator utilizing the intuitionistic fuzzy approach.

9.6 Implementation in MATLAB®

A MATLAB® code for image edge detection is given, which will be beneficial to the readers to implement the method.

An example to find the gradient image using dilation–erosion is given where Hamacher *t*-norm and *t*-conorms are used in the erosion–dilation computation:

```
clear all
image= imread('cameraman.tif');
dim1=240;dim2=dim1;row=dim1;dim=dim1;
 img1=imcrop(image,[1 1 dim-1 dim-1]);
figure,imshow(img1)
fil=fspecial('gaussian',5,0.6);
 filimg=uint8(filter2(fil,img1));
 img=double(img1);
mx=max(max(img));
mem1=img./mx;
newim = mem1;
 p1=0.2;p=0.8;
 c = [0,p,0;p,1,p;0,p,0];  % structuring element
 dim11=dim1+2;dim21=dim2+2;
 arow1=zeros(1,dim);
 w1=[arow1;newim;arow1];
 acol2=ones(dim+2,1);
 acol1=zeros(dim+2,1);
 fd2=[acol1,acol1,w1];
 fe3=[acol2,w1,acol2];
 lambda=3;
% performing erosion - dilation operation
for  j=2:dim2+1
 for i=2:dim1+1
   imd=[fd2(i-1,j-1),fd2(i,j-1),fd2(i+1,j-1);fd2(i-1,j),fd2(i,j),...
     fd2(i+1,j);fd2(i-1,j+1),fd2(i,j+1),fd2(i+1,j+1)];
```

```
          ime=[fe3(i-1,j-1),fe3(i,j-1),fe3(i+1,j-1);fe3(i-1,j),fe3(i,j),...
          fe3(i+1,j);fe3(i-1,j+1),fe3(i,j+1),fe3(i+1,j+1)];
       te2= (ime+c+(lambda-2)*ime.*c)./(1+(ime.*c)*(lambda-1));
       eros= min(min(te2));
       td1= (imd.*c)./(lambda+(imd+c-imd.*c*(1-lambda)));
       dil= max(max(td1));
       morphim_eros(i-1,j-1)=eros;
       morphim_dil(i-1,j-1)=dil;
  end
end
  mxdil=max(max(morphim_dil));
  morphim_dil1=255*morphim_dil./mxdil;
  mxeros=max(max(morphim_eros));
  morphim_eros1=255*morphim_eros./mxeros;
  edge= abs(uint8(morphim_dil1)-uint8(morphim_eros1));
figure,imshow(uint8(edge))    % gradient image
% finding binary edge
For i=1:dim1
  For j=1:dim2
       if edge(i,j)>50
       bin_edge (i,j)=1;
       else bon_edge(i,j)=0.0;
       end
     end
     end
figure, imshow(bin_edge);
```

9.7 Summary

This chapter describes morphology in general and fuzzy mathematical morphology in particular. Different fuzzy morphological operators given by Sinha and Dougherty, De Baets and Bloch are discussed. Also, fuzzy morphology, based on *t*-norms and *t*-conorms, and Lukasiewicz operator are discussed. Dilation and erosion are explained with examples. Application of fuzzy morphology in medical imaging particularly in edge detection along with the MATLAB program is also included.

References

1. De Baets, B. and Kerre, E., An introduction to fuzzy mathematical morphology, in *Proc. of North America Fuzzy Information Processing Society (NAFIPS'93)*, Allentown, PA, pp. 129–133, 1993.
2. Bandler, W. and Kohout, I.J., Semantics of implication operator and fuzzy relational products, *International Journal of Man Machine Studies*, 12, 89–116, 1980.

3. Bloch, I. and Maitre, H., Fuzzy mathematical morphologies: A comparative study, *Pattern Recognition*, 28(9), 1341–1387, 1995.
4. Bouchet, A., Pastore, J., and Ballarin, V., Segmentation of medical images using fuzzy mathematical morphology, *JCS&T*, 7(3), 256–262, 2007.
5. Burillo, P., Frago, N., and Fuentes, R., Fuzzy morphological operators in image processing, *Mathware & Soft Computing*, 10, 85–100, 2003.
6. Chaira, T., A novel intuitionistic fuzzy c means clustering algorithm and its application to medical images, *Applied Soft Computing*, 11(2), 1711–1717, 2011.
7. Deng, T.-Q. and Heijmans, H.J.A.M., Grey-scale morphology based on fuzzy logic, CWI Report PNA-R0012, Amsterdam, the Netherlands, October 2000.
8. Gasteratos, A., Andreadis, I., and Tsalides, P., Fuzzy soft mathematical morphology, *IEE Proceedings: Vision, Image and Signal Processing*, 145(1), 40–49, 1998.
9. Goecgherian, V., From binary to grey tone image processing using fuzzy logic concepts, *Pattern Recognition*, 12(1), 7–15, 1980.
10. González-Hidalgo, H., Massanet, S., and Torrens, J., Image analysis applications of morphological operators based on uninorms, in *Proc. of IFSA-EUSFLAT*, 630–635, Lisbon, Portugal, 2009.
11. Hamacher, H., *Über logische Aggregation nicht-binär explizierter Entscheidnungskriterien*, R.G. Fischer Verlag, Frankfurt, 1978.
12. Htun, Y.Y. and Aye, K.K., Fuzzy mathematical morphology approach to image processing, in *Proc. of World Academy of Science and Engineering*, vol. 44, p. 776, 2008.
13. Jiang, J.A. et al., Mathematical-morphology-based edge detectors for detection of thin edges in low-contrast regions, *IET Image Processing*, 1(3), 269–277, 2007.
14. Pahsa, A., Morphological image processing using fuzzy logic, *Havacilik Ve Uzay Knolojileri Dergisi, Ocak Cilt 2 Sayi*, 3, 27–34, 2006.
15. Popov, A.T., Interval based morphology color image processing, in Boyanov, T. et al. (eds.), *Lecture Notes in Computer Science*, Vol. 4310, Springer-Verlag Berlin Heidelberg, pp. 337–344, 2007.
16. Sinha, D. and Dougherty, E.R., An intrinsic approach to fuzzy mathematical morphology, in *SPIE*, Vol. 1607, Boston, MA, 1991.
17. Sinha, D. and Dougherty, E.R., Fuzzy mathematical morphology, *Journal of Visual Communication and Representation*, 3(3), 286–303, 1996.
18. Sinha, D. and Dougherty, E.R., Fuzzification of set inclusion: Theory and application, *Fuzzy Sets and Systems*, 55, 15–42, 1993.
19. Zhao, Y.-Q. et al., Medical images edge detection based on mathematical morphology, in *Proc. of IEEE Engineering in Medicine and Biology*, 6492–6595, Shanghai, China, 2005.

Index

A

Agglomerative clustering, 146
Associativity, 42

B

Bandler *t*-operators, 201
Boundary-based method, edge
    detection, 173–174
Boundary detection, 42
    boundary-based method, 173
    medical image processing, 24

C

Cauchy distribution, 132–136
Chow method, 113–114
CIELab colour space, 164–165, 186
Clustering, 34–35
    agglomerative, 146
    colour, 163–166
    dendrogram, 146
    divisive hierarchical clustering, 147
    FCM, 143–145
    fuzzy methods, 116–117
    hierarchical, 146–147
    IFCM
        Chaira algorithm, 157–159
        Iakovidis algorithm, 155–157
    kernel-based intuitionistic fuzzy
        clustering, 159–163
    kernel methods
        Ahmed's method, 153–155
        distance function, 148
        non-linear mapping function, 147
        objective function, 152–153
        radial basis function, 148
        robust FCM algorithm, 149–152
    non-fuzzy, 143
    type II fuzzy clustering, 167–168
Colour clustering, 163–166
Commutativity, 42
Content-based retrieval system, 28–29

Contrast enhancement
    by Chaira, 94–96
    fuzzy expected value, 87–88
    fuzzy histogram hyperbolization, 86
    grey-level mapping, 84
    IF-THEN rules, 86
    intensification operator, 86
    low intensity areas, 83
    transformation function, 85

D

Decision-making process, retrieval
    system, 29
Dilation
    Lukasiewicz operator, 199
    morphological processing
        operations, 193–194
Distance measure, 63, 65
    Hausdorff metric, 66–67
    IVIFS, 79–80
    Song and Zhou measure, 67
    Szmidt and Kacprzyk measure, 66
    Wang and Xin measure, 67
Divergence
    fuzzy methods, 114–115
    intuitionistic fuzzy methods, 121–125
Divisive hierarchical clustering, 147

E

Edge detection, 35–36
    accurate technique, 184–188
    boundary-based method, 173–174
    fuzzy mathematical morphology,
        207–208
    fuzzy methods, 174–175
    gradient-based methods, 171
    Hough transform method, 173
    interval-valued fuzzy relation,
        181–183
    intuitionistic fuzzy edge detection
        method, 175
    Laplacian method, 171

MATLAB®
   edge image, 188–190
   fuzzy edge image, 190–191
   median filter, 178–180
   smoothing, 172
   template-based edge
     detection, 176–178
   thresholding method, 172–173
   type II fuzzy set, 183–184
Edge image, MATLAB, 188–190
Entropy-based methods, 88
   by Burillo and Bustince, 89–90
   by Chaira, 93–94
   fuzzy method, 174–175
   intuitionistic fuzzy
     methods, 118–121
   linear index of fuzziness, 90
   maximization index of fuzziness,
     90–91
   by Vlachos and Sergiadis, 89
Entropy measure, 63
   IFE, 74–78
   interval-valued intuitionistic fuzzy
     set, 78–79
Erosion
   Lukasiewicz operator, 199
   morphological processing
     operations, 193–194

F

FCM clustering, *see* Fuzzy c means
     (FCM) clustering
Feature detection, 27
Feature matching, 27
FEV, *see* Fuzzy expected value (FEV)
Footprint of uncertainty (FOU), 18
Fuzzy aggregating operators, 47
   ordered weighted averaging
     operator, 48
   weighted averaging operator, 48
Fuzzy c means (FCM) clustering, 27,
     143–145
Fuzzy complement, 5–8
Fuzzy divergence-based edge
     detector, 175
Fuzzy edge detection, 174–175
   median-based edge detector, 175
   Sobel edge detector, 174

Fuzzy edge image, MATLAB, 190–191
Fuzzy expected value (FEV), 87
Fuzzy histogram hyperbolization, 86
Fuzzy image processing, 29–31
Fuzzy mathematical morphology
   binary morphology, 195
   by Bloch and Maitre, 197
   by De Baets and Kerre, 197
   edge detection, 207–208
   fuzzy conjunction, 196
   greyscale, 194–195
   Hamacher operation, 196
   intuitionistic fuzzy approach,
     208–211
   Lukasiewicz operation, 196, 198–200
   MATLAB, 211–212
   opening and closing operations,
     202–206
   preliminaries, 193–194
   by Sinha and Doughert, 198
   t-norms and *t*-conorms, 200–202
Fuzzy median-based edge
     detector, 175
Fuzzy methods
   clustering, 116–117
   divergence, 114–115
   geometry, 115
Fuzzy operators
   decision-making problem, 40–41
   logic operators, 40
   multi-valued logic, 39
   negation, 47
   t-conorm, 42–47
   t-norm, 41–42
Fuzzy Sobel edge detector, 174

G

Generalized intuitionistic fuzzy hybrid
     averaging (GIFHA) operator,
     54–57
Generalized intuitionistic fuzzy
     ordered weighted averaging
     (GIFOWA) operator, 52–54
Generalized intuitionistic fuzzy
     weighted averaging (GIFWA)
     operator, 49–52
Geometry, fuzzy methods, 115
Global thresholding, 109–111

Gradient-based methods, edge
  detection, 171
Greyscale mathematical morphology,
  194–195

H

Hamacher *t*-conorm
  image enhancement, 100
  MATLAB, 105–106
Hamacher *t*-operators, 201–202
Hausdorff metric, 66–67
Hesitancy histogram equalization,
  96–99
Hesitation degree, 2, 120
  intuitionistic fuzzy divergence
    measure, 123–124
  membership value, 127
Hierarchical clustering, 146–147
Hough transform method, edge
  detection, 173
Hysteresis thresholding, edge detection,
  172–173

I

IFCM clustering, *see* Intuitionistic fuzzy
  c means (IFCM) clustering
IFD, *see* Intuitionistic fuzzy
  divergence (IFD)
IFE, *see* Intuitionistic fuzzy
  entropy (IFE)
IFRs, *see* Intuitionistic fuzzy
  relations (IFRs)
IFS, *see* Intuitionistic fuzzy set (IFS)
IF-THEN rules, contrast enhancement
  using, 86
Image enhancement, 83
  contrast enhancement, 23, 33, 84–85
  by Chaira, 94–96
  fuzzy expected value, 87–88
  fuzzy histogram
    hyperbolization, 86
  IF-THEN rules, 86
  intensification operator, 86
  Hamacher *t*-conorm, 100
  MATLAB, 102–107
  type II fuzzy set, 99–102

Image fusion, 28
Image registration
  atlas registration, 27
  dimensionality, 25
  domains, 26
  goals, 24
  interaction, 26
  inter-subject registration, 26
  methods, 27
  modalities involved in, 26
  natures, 25–26
  optimization procedure, 26
  stereotactic frame, 25
  subject, 26–27
Image retrieval, 28–29
Image segmentation, 24, 34
Image transformation, 27
Intensification operator, 86
Interval-valued fuzzy relation, 181–183
Interval-valued intuitionistic fuzzy set
  (IVIFS), 16–17, 78–80
Intuitionistic Euclidean distance, 66
Intuitionistic fuzzy approach, 208–211
Intuitionistic fuzzy c means (IFCM)
    clustering
  Chaira algorithm, 157–159
  Iakovidis algorithm, 155–157
Intuitionistic fuzzy divergence (IFD),
  70–72, 123–125
Intuitionistic fuzzy edge detection
    method, 175
Intuitionistic fuzzy enhancement
    methods
  entropy-based enhancement
    methods, 88–92
    by Burillo and Bustince, 89–90
    by Chaira, 93–94
    by Vlachos and Sergiadis, 89
  hesitancy histogram equalization,
    96–99
  2D entropy-based IF enhancement,
    92–93
Intuitionistic fuzzy entropy (IFE), 74
  Chaira, 75
  graphical representation, 76
  Huang and Liu, 78
  Kacprzyk, 78
  Szmidt and Kacprzyk, 75–77
  Vlachos and Sergiadis, 78

Intuitionistic fuzzy generator
   Chaira-type, 10
   fuzzy complement and, 5–8
   non-membership values, 8
   Sugeno-type, 9
   Yager-type, 9
Intuitionistic fuzzy index, 2
Intuitionistic fuzzy measure,
       70–72, 117–118
   divergence-based method, 121–125
   entropy-based method, 118–121
   information measure, 73–74
Intuitionistic fuzzy relations
       (IFRs), 10
   reflexive property, 14–15
   supremum–infimum, 11
   symmetric property, 15
   t-conorm, 12–14
   t-norm, 12
   transitive property, 15–16
Intuitionistic fuzzy set (IFS)
   medical image processing, 31–32
   membership function, 1
   multi-attribute decision-making,
       57–58
   non-membership degree, 1
   operations on, 3–5
   representation, 2–3
   triangular conorms, 58–59
   weighted averaging operator, 48–49
      GIFHA operator, 54–57
      GIFOWA operator, 52–54
      GIFWA operator, 49–52
Intuitionistic Hamming distance, 66
Intuitionistic normalized Euclidean
   distance, 66
Intuitionistic windowed thresholding
   method, 136–138
Iterative thresholding, 111
IVIFS, *see* Interval-valued intuitionistic
   fuzzy set (IVIFS)

K

Kanenko method, 113–114
Karhunen–Loeve transform (KLT), 29
Kernel clustering methods
   Ahmed's method, 153–155
   distance function, 148

non-linear mapping function, 147
   objective function, 152–153
   radial basis function, 148
   robust FCM algorithm, 149–152
Kohout t-operators, 201

L

Laplacian method, edge detection, 171
Leucocyte images, blood cells,
      132–136
Locally adaptive thresholding, 111–113
Lukasiewicz t-operators, 202
Lukasiewicz union, 42

M

MATLAB
   edge detection
      edge image, 188–190
      fuzzy edge image, 190–191
   fuzzy clustering, 168
   fuzzy mathematical morphology,
      211–212
   Hamacher t-conorm, 105–106
   image enhancement, 102–107
   intuitionistic fuzzy-enhanced image,
      106–107
   intuitionistic windowed
      thresholding method, 136–138
   type II fuzzy
      image enhancement, 105–106
Median filter, edge detection using,
      178–180
Medical image processing
   applications, 33–36
   boundary detection, 24
   contrast enhancement, 23
   CT scan, 36
   fusion, 28
   fuzzy processing, 29–31
   intuitionistic fuzzy set, 31–32
   morphology, 24
   registration, 24–27
   retrieval, 28–29
   segmentation, 24
   type II fuzzy set, 32–33
Model transformation, 27
Monotonicity, 42

Morphology, 24, 36
Multispectral thresholding, 114

N

Nilpotent *t*-conorm, 42

O

Optimal thresholding, 111
Ordered weighted averaging (OWA)
  operator, 47–48

P

Product union, 42
P-tile thresholding, 110

S

Segmentation
  clustering (*see* Clustering)
  contextual technique, 109
  grouping, 143
  leucocyte images in blood cells,
    132–136
  medical image processing, 24
  non-contextual technique, 109
  type II fuzzy set theory, 131–132
Similarity measure, 63–65
  Dengfeng and Chuntian measure,
    68–69
  Hausdorff distance, 68
  Hung and Wang measure, 69
  IVIFS, 79–80
  Liang and Shi measure, 69–70
Smets–Magrez's axioms, 198
Sugeno fuzzy generator, 165
Sugeno's intuitionistic fuzzy
  generator, 185
Sugeno's λ complementation, 200
Sugeno-type fuzzy complement, 6, 8

T

T-conorm
  Chaira's, 44–46
  Dombi's, 46–47
  Dubois's, 44

Frank's, 46–47
fuzzy mathematical morphology
  using, 200–202
Hamacher's, 46
Prade's, 44
Schweizer's, 43
Sklar's, 43
Sugeno's, 44
Weber's, 43
Yager's, 43
Zadeh's, 43
Template-based edge detection,
  176–178
Text-based retrieval system,
  28–29
Thresholding method
  Chow and Kanenko method,
    113–114
  edge detection, 172–173
  fuzzy methods
    clustering, 116–117
    divergence, 114–115
    geometry, 115
  global, 109–111
  intuitionistic fuzzy
    methods, 117–118
    divergence-based method,
      121–125
    entropy-based method, 118–121
  iterative, 111
  locally adaptive, 111–113
  MATLAB
    intuitionistic windowed
      thresholding method, 136–138
  membership function calculation,
    126–128
  multispectral, 114
  optimal, 111
  type II fuzzy set theory, 129–130
  window-based thresholding,
    125–128
T-norm
  Chaira's, 44–46
  Dombi's, 46–47
  Dubois's, 44
  Frank's, 46–47
  fuzzy mathematical morphology,
    200–202
  Hamacher's, 46

Prade's, 44
Schweizer's, 43
Sklar's, 43
Sugeno's, 44
Weber's, 43
Yager's, 43
Zadeh's, 43
2D entropy-based IF enhancement,
    92–93
Type II fuzzy set, 1
    clustering, 167–168
    edge detection, 183–184
    image enhancement, 99–102
    interval-based, 19
    medical image processing,
        32–33
    segmentation, 131–132

thresholding, 129–130
uncertainty, 18

W

Weighted averaging operator, 48–49
    GIFHA operator, 54–57
    GIFOWA operator, 52–54
    GIFWA operator, 49–52
Window-based thresholding, 125–128

Z

Zadeh's concentration operation, 200
Zadeh's union, 42
Zadeh *t*-operators, 200–201
Zero identity, 42